Planning for Conflict in the Twenty-First Century

PRAEGER SECURITY INTERNATIONAL ADVISORY BOARD

BOARD COCHAIRS

Loch K. Johnson, Regents Professor of Public and International Affairs, School of Public and International Affairs, University of Georgia (U.S.A.)

Paul Wilkinson, Professor of International Relations and Chairman of the Advisory Board, Centre for the Study of Terrorism and Political Violence, University of St. Andrews (U.K.)

MEMBERS

Anthony H. Cordesman, Arleigh A. Burke Chair in Strategy, Center for Strategic and International Studies (U.S.A.)

Thérèse Delpech, Director of Strategic Affairs, Atomic Energy Commission, and Senior Research Fellow, CERI (Fondation Nationale des Sciences Politiques), Paris (France)

Sir Michael Howard, former Chichele Professor of the History of War and Regis Professor of Modern History, Oxford University, and Robert A. Lovett, Professor of Military and Naval History, Yale University (U.K.)

Lieutenant General Claudia J. Kennedy, USA (Ret.), former Deputy Chief of Staff for Intelligence, Department of the Army (U.S.A.)

Paul M. Kennedy, J. Richardson Dilworth Professor of History and Director, International Security Studies, Yale University (U.S.A.)

Robert J. O'Neill, former Chichele Professor of the History of War, All Souls College, Oxford University (Australia)

Shibley Telhami, Anwar Sadat Chair for Peace and Development, Department of Government and Politics, University of Maryland (U.S.A.)

Fareed Zakaria, Editor, Newsweek International (U.S.A.)

Planning for Conflict in the Twenty-First Century

Brian Hanley

Foreword by Jason Armagost

PRAEGER SECURITY INTERNATIONAL
Westport, Connecticut • London

Library of Congress Cataloging-in-Publication Data

Hanley, Brian, 1961–
  Planning for conflict in the twenty-first century / Brian Hanley.
    p. cm.
  Includes bibliographical references and index.
  ISBN 978-0-313-34555-5 (alk. paper)
 1. Military education—United States. 2. History—Study and teaching
(Higher)—United States. 3. Soldiers—Education, Non-military—United
States. 4. Military planning—United States. 5. Strategy. I. Title. II. Title:
Planning for conflict in the 21st century.
  U153.H36 2008
  355.0071'173—dc22   2007035415

British Library Cataloguing in Publication Data is available.

Copyright © 2008 by Brian Hanley

All rights reserved. No portion of this book may be
reproduced, by any process or technique, without the
express written consent of the publisher.

Library of Congress Catalog Card Number: 2007035415
ISBN-13: 978-0-313-34555-5

First published in 2008

Praeger Security International, 88 Post Road West, Westport, CT 06881
An imprint of Greenwood Publishing Group, Inc.
www.praeger.com

Printed in the United States of America

The paper used in this book complies with the
Permanent Paper Standard issued by the National
Information Standards Organization (Z39.48-1984).

10 9 8 7 6 5 4 3 2 1

Contents

Foreword by Jason Armagost		vii
Acknowledgments		xi
Introduction		xiii
Chapter 1	Lessons Not Learned: Strategy, War Plans, and the United States Armed Forces	1
Chapter 2	Transformation Ballyhoo	58
Chapter 3	The Brain of an Army: Establishing a First-Rate Joint War University	84
Chapter 4	From the Ardennes to Dunkirk: France 1940	111
Chapter 5	Stalingrad	145
Chapter 6	Desert War: From El Alamein to Cape Bon	169
Concluding Thoughts		185
Appendix		187
Endnotes		191
Works Consulted		201
Index		211

Foreword

Every end does not appear together with its beginning.

Herodotus, *The Histories*

Herodotus: Piloting a Stealth Bomber bound for Baghdad, I recalled his magnificent narrative of the wars between Greek and Persian. Opening night of *Iraqi Freedom*, 18 hours into the mission, eastbound from Missouri to Iraq; another 20 hours until our wheels would touch down at the home airfield, and I couldn't shake from my mind *The Histories*. I recalled Herodotus's digressions on the deadly flying snakes of Arabia, the Amazons who must kill a man before they can marry, the temple prostitutes of Babylon, and the fascinating excursus on Egypt. Herodotus is no mere fabulist, however. His work complements Homer's epic poetry and also sets the stage for the Attic literature that bequeathed to us the mythogenic struggles at Marathon, Thermopylae, and Salamis.

In the midst of reflections on ancient history—and with the horizon lit by AAA and SAM fire—I released one 5,000-pound GPS-guided "Bunker Buster." A trio of one-ton high-explosive penetrators followed. These weapons laid waste to a command center in the basement of Saddam's main Presidential Palace—bringing to mind a lightning bolt hurled by Zeus. Broadcast live on 24-hour news channels, this moment was the culmination of 15 months of preposterously rigorous planning, coordination, and jargon-festooned briefings.

Despite my private inclination to view technology with other than idolatrous eyes, on the mission to Baghdad I found myself engrossed in the operation of computers, broadband satellite radios, digital engine controls, and guided weapons. I owe my physical safety and combat effectiveness to the wizardry of stealth technology, but my warrior soul is by no means instructed by superheterodyne, nuclear-hardened processors, and blended-wing composite airframes. The gadgetry and processes—while often decisive in tactical engagements—can lead us to forget that war is first and foremost a struggle of will and mind, the most intense form of social interaction. This timeless truth has been brilliantly cast in *Planning for Conflict in the Twenty-first Century*. The book's argument is cogent, grave, and majestic, embellished by ripping good humor and perceptive historical analysis. Hanley's narrative strides through new and familiar terrain with a zest that is as pleasing as it is illuminating.

I find Hanley's characterization of planning staffs and the officers who command these organizations especially engaging because Hanley is a gifted observer whose commentary deserves attention. Every system is perfectly designed for the product it begets. This holds true for the mind and character of the modern flag officer and strategist. No one would mistake any flag officer I have ever encountered for a dunderhead, but too often such an officer's ambition centers on paying homage to, and defending, the bureaucratic system that has feathered his or her cap. The desire to get ahead, to make rank, attenuates or befogs one's understanding of officership in the ideal. The most successful officers (based on the nominal qualifier of rank) are besotted with trendy enthusiasms derived from the sociological view of the MBA student. Professor Hanley dares to demand more from our leaders. He poses and answers decisively the question of how to educate officers in ways that increase the possibility that the U.S. Armed Forces will produce senior leaders endowed with the well-tutored and supple intellect of a Washington, Bismarck, Churchill, or Liddell-Hart.

If our planning staffs are to be damned for one thing in particular, it is that, in their hurry and bustle, they provide answers to the wrong questions. Their lodestar is a meretricious compound of intellectually arid doctrine and catchphrases destitute of substantive meaning. Planning becomes an exercise in discovering problems that conform to ready-made solutions. In taking on what really is an intellectually friable status quo, Hanley strikes a concussive blow. Witness his commentary on the disheartening absence of a clear-headed strategic plan for transforming Iraq into a reliable ally—cultural as well as strategic—of the United States.

We ignore at our peril the core of Hanley's argument that a narrow ration of first-hand combat experience seasoned by bureaucratic sophistry can never compensate for the absence of historical knowledge fortified by a mind properly trained to wield it. As an active duty officer, I am indebted to Professor Hanley for the time we've spent discussing wars and generalship, past and present. I am honored by the opportunity to write a foreword to this splendid book, which is at

Foreword

heart a work of reclamation. Hanley is a gadfly, an iconoclast, a stalk-and-ambush slayer of sacred cows. He challenges our basic assumptions about the planning and conduct of modern war in ways that should make those of us with recent combat experience grateful.

Hanley's book is erudite and jocular, with a classicist's sense of irony. Any citizen who cares about the collective mind of our military leaders will profit by reading and rereading *Planning for Conflict in the Twenty-first Century*, which serves as an indispensable resource for us to rethink the nature and purpose of the education of American warriors and statesmen. I thank him for doing work that is every bit as noble as it is difficult.

As one begins with Herodotus, so one must end:

> *He is the best man who, when making his plans, fears and reflects on everything that can happen to him, but in the moment of action is bold.*

Lieutenant Colonel Jason Armagost, USAF
August 2007
Sedalia, Missouri

Acknowledgments

I am indebted to a number of people whose advice, encouragement, and constructive criticism made this book possible: Lt Col Jason Armagost (USAF), Lt James Dolbow (USCGR), Professor Jeanne Heidler (USAF Academy), Lt Col Jim Lacey (AUS, Ret.), Professor James H. Meredith, Lt Gen James Mattis (USMC), and MCPOCG (Ret.) Vincent Patton III. Luck—or Providence—brought me into contact with the United States Naval Institute in 2005. I have benefited immensely from my association with Major General Tom Wilkerson (USMC, Ret.), CEO of the United States Naval Institute, the staff of the United States Naval Institute, and my colleagues on the Institute's Editorial Board of Directors—all of whom breathe life into the Institute's motto: "Transforming Defense through the power of ideas." My students at the Joint Forces Staff College were an inspiration to me as well, particularly Lt Col Dave Tabor (USAF) and Captain Tom McDonough (USN). Most important, I am indebted to my wife, Terry, and my son, Bryce, whose love and encouragement sustained me throughout the research and writing of this book.

Introduction

This book is part essay and part memoir. Twenty years experience in the United States Air Force (USAF) has informed my views to a considerable extent, but the greater impetus behind this work came from a lifetime's reading in military history. The eleven-year-old boy I remember was enthralled by the historical writings that adorned the shelves not only at home and in bookstores, but also the magazine rack at a nearby shop. My father, a nose gunner in a B-24 Liberator during the Second World War, would drive me to the neighborhood pharmacy where, once a week or so, I would be allowed to pick up one of the monographs from the *Ballantine Illustrated History of World II* series and, if he were in an unusually generous mood, a styrene plastic model kit (my favorites were battleships and tanks—upon which I inevitably imposed a personal signature, glue-stained fingerprints). The Ballantine editors also published a monthly magazine on World War II. After scrutinizing each issue soon after it arrived, I would dutifully file it away in one of the gold-embossed black binders that came with the subscription.

Reading the Ballantine books was for me something more than a pleasant diversion. I recall enjoying dinner out with what my father, at the time, remarked was an unseemly and greedy relish, my appetite having been abnormally stimulated earlier in the afternoon by reading Alan Wykes' account of the death-dealing privation imposed on the residents of Leningrad by German forces that had invested the city. It would be false to say that the Ballantine authors made history come alive. History cannot help but live, and it doesn't need to be resuscitated or resurrected, if only because the present is a seamless outgrowth of the past. Though they hardly qualify as masterworks, the Ballantine books exemplify good historical writing. In only 160 pages—many of which were given over to photographs and drawings—the author of each volume provided a thoroughly engaging discussion of a given subject: battles, campaigns, weaponry, and commanders. It is very much to be regretted that the series is long out of print and that the paperback format—unlike the lucidity of the writing and the cogency of the argument—does not stand up well to the passage of time.

Nothing like the Ballantine series exists today, and even if there were such a collection in print one might wonder who would buy it? The rising generation would hardly be expected to constitute a readership. It is likely that the elegant prose would soon overtax the patience of contemporary adolescents and young adults—who, if they have a fancy at all to learn about the battle for Tarawa or the Raid on Saint-Nazaire, can pick up the television remote control and catch a twenty-minute video presentation. This is a pity. History rightly understood is a narrative conveyed by words on a page, the logic of which requires an understanding of the complex, at times contradictory, and perhaps unfathomable motivations of many people acting on circumstance. History can be neither expressed nor understood through a medium best suited to lowbrow entertainment and placed in the service of retailers peddling their wares—potato chips, automobiles, drugs—that claim to have eclipsed character and education in realizing contentment. It is true that there are plenty of history books published nowadays—clothbound, adorned with footnotes, and comprising many hundreds of pages, their authors impressively credentialed—but few if any of these can inspire in the same way that the Ballantine books nourished my imagination and impelled me toward a career in the armed forces.

Having spent more than twenty years on active duty has taught me the foolishness of placing great faith in experience that is divorced from a cultivated historical understanding. In reflecting on my time in the USAF, one realizes that a great majority of the day is spent discharging humdrum obligations—mastering one's technical specialty, keeping abreast of administrative matters, wearing out the periods of inactivity on deployments with shoptalk and other strains of light banter—while the duty to study one's profession, and a duty it must always be, declines in status to an exotic and dispensable pastime. What happens to most of us is that we become increasingly absorbed in our careers, and gradually the weight of obligation and habit produces an efficient administrator or technician utterly bereft of intellectual resourcefulness and allergic to rigorous, independent thought.

At first glance this frame of mind might appear unexceptionable—service members are expected to skillfully operate the bureaucratic and technical machinery of the modern military—but for officers who by rank or position are expected to exercise sound judgment on strategic and operational matters it can be dangerous. Commanders and war planners most especially should unceasingly develop their critical intelligence, or at least keep it from ossifying in the sterile atmosphere of bureaucratic culture. It is a misfortune that the outlook of the technician and administrator has come to dominate our profession. Our doctrine and our attitude to professional education have enshrined the idea that war is largely a mechanical activity rather than an intense form of social intercourse—a state of affairs that can only disquiet the student of military history.

I was moved to write this book in part by my experiences as a faculty member at the Joint Forces Staff College (JFSC), Norfolk, VA. The underlying weakness of the Joint & Combined Warfighting School, which handles the bulk of the college's students, is that the curriculum is based on the terribly misbegotten idea that an officer can be prepared for service on a joint staff—which among other duties produces war plans—by taking a course of a few weeks duration that focuses entirely on administrative matters. I was able to observe firsthand the establishment of the Joint Advanced Warfighting School (JAWS), which was intended to offer a postgraduate education in war planning. Sadly, JAWS turned out to be nothing more than a gargantuan replica of the shorter course, with little prospect of improvement given the entrenched culture of the institution. The Joint Forces Staff College—which can only reflect habits of mind that obtain throughout the Department of Defense (DOD)—is devoted to producing bureaucratic minions, rather than strengthening the intellectual ability of officers who will help write the war plans upon which our national security ultimately depends.

The old saw that war is too important to be left to generals is, I think, foolish because it is unhistorical; it is akin to asserting that the practice of medicine and law are too important to be left to physicians and attorneys. But as officership deteriorates into a synonym for "technician/administrator," and thus grows estranged from its traditional connotation "resolution in the service of sound judgment," the canard that military officers cannot be trusted to wield ideas might yet prove to be true. The present volume will have served the military profession, and so the national interest, if it helps to generate reform on how the armed forces train, educate, and promote officers who shape our military strategy and write our war plans. Collaterally, my hope is that military readers in particular will discover the professional and intellectual improvement that wide reading in the masters of historical narrative offers to them.

March 2007
Kissing Camels Estates
Colorado Springs, CO

Chapter 1

Lessons Not Learned: Strategy, War Plans, and the United States Armed Forces

The most influential official document of the twentieth century came not from the desk of a head of state, but from the hand of an officer in charge of a military planning staff. The aims and assumptions of the Schlieffen Plan reflected the vainglory of European civilization, as well as the antagonisms and rivalries that in maturity would undermine a culture more than 500 years in the making. War is the single greatest actuator of human progress and misery, and one cannot understand the twentieth century—culturally, politically, demographically—without grasping the causes of the First World War and its aftermath. War plans determine whom we fight, how we fight, where and under what provocation we fight—and to a large extent they shape the peace that follows. All plans take into account the uncertainty of war, but those that are judiciously conceived limit the unattractive options that confront civilian and military decision makers once the shooting starts.

In the American mind, the establishment of a general staff devoted to building war plans in the abstract has always been a faintly disreputable idea. There are a number of reasons for what seems to be an implausible cultural trait, given the supremacy of our military might at the beginning of the twenty-first century. Like most every other nation, the United States was established by war, which in our case was less a revolution than an act of secession realized by violence. But unique among major powers in the Western world, the founders of the United States quite consciously eschewed anchoring their legitimacy in military prowess, either their own or that of their ancestors. The

United States was and perhaps still is seen as a grand experiment: a country founded on rights that inhere in the individual, and that is devoted to free enterprise. The ideas—the perfectibility of man, the decadence of existing social and political orders in Europe, the primacy of reason and experience in explaining phenomenon—generated by Enlightenment thinkers such as Voltaire, Adam Smith, Rousseau, David Hume, and the Encyclopedists shaped the mental outlook of the author of the *Declaration of Independence* most especially.

The first commander in chief of the American army accepted the appointment not with an eagerness born of an appetite for martial glory, but with the modest reluctance of one who sees military service as a civic obligation. Washington's military reputation was built on his achievements in the service of the Crown during the French and Indian War, but he thought of himself not as a professional soldier, but as a gentleman who made his living as a planter. "Tho' I am truly sensible of the high Honour done me in this Appointment," George Washington wrote to the President of the Continental Congress (16 June 1775):

> yet I feel great distress from a consciousness that my abilities and Military experience may not be equal to the extensive and important Trust: However, as the congress desires I will enter upon this momentous duty, and exert every power I Possess In their Service for the Support of the glorious Cause: I beg they will accept my most cordial thanks for the distinguished testimony of their Approbation. But lest some unlucky event should happen unfavourable to my reputation, I beg it may be remembered by every Gentn. in the room, that I this day declare with utmost sincerity, I do not think myself equal to the Command I am honoured with.[1]

How different in character and temperament was Washington from the French general who midwifed modern France!

American indifference to martial culture was not lost on distinguished observers from abroad. Alexis de Tocqueville says nothing about the armed forces of the United States, apart from making a prediction that the growth of commerce will require the development of a first-rate navy. Seven decades after the publication of *Democracy in America*, James Bryce observed much the same thing. In the fifteen hundred pages that comprise *The American Commonwealth*, only a few paragraphs are given over to commentary on American military power—and on this point Bryce can be said to echo de Tocqueville in his remark that a unique feature of American life is that a lively patriotism coexists with a diminutive military force and a desire to avoid war. From the start, then, Americans have viewed war as an aberration—neither a defining part of our past nor a necessary or inevitable constituent of our future. Naturally enough, the idea of planning for wars that might never come to pass has been at odds with our national character.²

Also contributing to America's aversion to the founding of a general staff devoted to war planning is our British legal heritage. Standing armies were held to be the instrument of tyrants. When the last English king who asserted royal prerogatives was put to death in 1649, he was replaced by Oliver Cromwell—a general who, on the way to establishing a despotism of his own, employed his new model army to dispense with the legislature he professed to serve. The Continental Congress, for example, steadfastly rebuffed General Washington's plea for the establishment of an office that would handle ordinary staff duties, on the grounds that doing so would give too much power to a military commander. And even after a "Board of War" was created, the Continental Congress controlled most of its functions.³

The first seven decades of our existence were relatively peaceful when compared with the European experience over the preceding two centuries. The Civil War was, for us, an anomaly. On the one side, an inadequately provisioned force that was expertly led, fought with great skill to achieve a limited strategic objective; on the other side, a large and well-provisioned army, comprised, for the most part, of volunteers, but equally endowed with talented commanders (late blooming though they were), succeeded in bludgeoning its opponent into submission. In the years following Appomattox, the armed forces of the United States fell into the mold that had been cast for them at the nation's founding. The chief duty of the army was to put down the belligerence of Indian tribes, quell domestic disturbances (usually the result of labor disputes), skirmish with enfeebled rivals such as Spain and Mexico who sought to challenge Manifest Destiny, and pacify a handful of petty colonial territories in the Caribbean and the Pacific. The navy—of a suitable size by 1900 and, unlike the army, thoroughly modernized—guarded our commercial shipping routes and kept watch over our coasts.

In spite of the immense achievements of the Prussian General Staff in the war with France 1870–1871, indifference toward operational planning obtained among America's senior military leadership through the first years of the twentieth century. Up until that time the United States made no serious effort to plan for war in any systematic way. The usual practice was for the president to seek a declaration of war from Congress, followed by a rapid expansion of the ranks by volunteers. In the mean time the president and his advisers would devise strategic objectives. Thanks in large measure to Alfred Thayer Mahan, in the early 1890s the Navy War College began drafting war plans that reflected America's standing as a newly hatched world power, but these drew no connections between operational and strategic objectives, and—of much greater importance—nothing was said about the peace sought by the other side. One plan, for example, was based on war with Great Britain, but it took no account of what might provoke the conflict, nor what political and diplomatic aims were to be achieved, and not much thought was given to collaboration with the army. At about the same time, the army began studying the possibility of British aggression against the United States. The notional plan that army officers came up with centered on an invasion of Canada. The navy was not consulted.[4] In spite of their flaws, these exercises were wholly commendable, but they reflected neither a mature understanding of war planning (as exemplified by the work of the German General Staff), nor the priorities of the War Department.

In expressing his opposition to Elihu Root's move to create a general staff in 1902, General Nelson A. Miles, chief of staff of the army, asserted that good generalship obviates the need for detailed planning. "As far as a plan of campaign is concerned," General Miles declared during congressional hearings on the subject, "that must depend on circumstances, and if a general is not able to make a plan and carry it out *instantly*, he is not competent to command an army, or a division, or a corps."[5] General Miles was no doubt motivated by parochial and self-serving concerns: Root's General Staff idea threatened the authority of the office of chief of staff of the army. Even so, the intellectual underpinnings of Miles' argument reflected America's limited experience with modern war. It is unlikely that General Miles or any other influential officer had read *On War*, though Clausewitz's ideas were critically evaluated in Jomini's *Summary of the Art of War*, a book that was widely known in the United States in the nineteenth century and provided the theoretical basis for such military doctrines as existed. An English translation of *On War* appeared in 1874, but there is little evidence to suggest the book was much read on the western side of the Atlantic. Even so, Miles's viewpoint can be said to reflect the distinctively romantic notion of genius that Clausewitz adapted to the profession of arms in what is perhaps the most engrossing section of his book. It is far more likely that General Miles' perspective was informed by indifference or a misunderstanding of recent German victories over Austria and France, and fortified by an uncritical appreciation of Napoleon, Charles XII, Wellington, Gustaphus Adolphus, and Julius Caesar. These legendary commanders triumphed because of their robust battlefield intuition and the élan of their troops. Operational planning for them was never much more than an ad hoc consideration of transport and logistics. Before the age of Helmuth von Moltke, up-and-coming field commanders most often were tutored by brilliant and charismatic patrons; deliberate war planning was never put in the form of a system.[6]

A word must be said about Elihu Root (1845–1937), who served as the United States secretary of war from 1899 to 1904. Apart from a brief stint in the New York militia, Root never served in the armed forces. When President McKinley asked him to replace the feckless Russell A. Alger as secretary of war, Root suspended his flourishing legal practice in New York and uprooted his family, not to pursue fame or fortune, but out of a highly developed civic-mindedness, and probably also because he assumed his tenure would be brief—a refreshing sabbatical from the familiar routines of law. Perhaps because he was untouched by the traditions and crotchets of military culture, Root brought intellectual vibrancy, enormous energy, and an impermeable disinterestedness to his job, which allowed him to see clearly the inadequacies of the status quo. He possessed a keen apprehension of both the flaws in contemporary military culture and the specific remedies that would most effectively address them. Root imposed systematic operational planning on a United States Army that had at least mildly resisted the idea up until that time and for a short while afterward. His achievement stands as a rebuke to the habits of mind that bureaucracies often encourage: dim-witted inertia born of careerism, reflexive contempt for ambitious ideas, and truculently parochial self-seeking.[7]

Root was also endowed with a prescient understanding of America's strategic circumstances. The extirpation of Spain from the Western Hemisphere and our acquisition of her overseas possessions, in combination with our astonishing industrial and demographic growth, meant that the United States had become a world power whose exalted position would bring forth competitors and enemies as well as allies. Root also appreciated the ramifications of technological advances on military operations. No commander could ever again enjoy the luxury of a panoramic view of the battle, as had been the case up until the first half of the nineteenth century. The rapid mobilization of vast armies comprised mostly of conscripts and reservists; the extended range and accuracy of artillery; the development of the magazine-loading rifle; the steam engine; the advent of the machine gun, the telegraph, and the field telephone: these things would make command in the field infinitely more difficult, and perhaps impossible, were not an efficient staff system in place to invigorate and discipline the creation, transmission, and execution of military orders. Root understood that the great warlord approach to fighting— in which an illustrious commander, sometimes by his mere presence, determined the outcome—must yield to the demands of administration and the ineluctable industrialization of modern combat.

Incredible though it may seem given the work of Elihu Root, the United States Army did not begin working in earnest on what we would nowadays consider deliberate war planning, until the middle of President Woodrow Wilson's first term, which was shortly before the outbreak of war in Europe. War Plan Orange dealt with the possibility of Japanese aggression against the Philippines. War Plan Black assumed an attempt by Germany to dominate the Western Hemisphere by invading the United States directly, or by establishing a base of operations in the West Indies.

Even when war broke out in Europe, updated editions of Plan Black and Plan Orange reflected our isolationist temper more than strategic reality. In early 1917 the United States had drawn up plans based on the possibility of a Japanese invasion of the West Coast. The plan for Europe envisioned an American expeditionary force launching an offensive against Bulgaria via Greece, and attacking German forces in France in concert with the Dutch army. No serious consideration was given to fighting as part of a coalition that included the French and British armies. Much of the blame for this rather pathetic state of affairs must be laid at the feet of President Wilson, who preferred moral preening in the guise of diplomacy and only reluctantly asked for a declaration of war, after the wide publication of the Zimmerman telegram and the German decision to unleash unrestricted submarine warfare forced him to do so. The armed forces deserve censure as well, if only because the plans endorsed by the senior leadership took no account of the character and likely trajectory of the war then in progress.[8]

The outlook of the manager and the technician had come to dominate American culture by the mid-1920s, and this, combined with our irenic temper, gave license to—indeed conferred a special authority on—a mechanical approach to war planning.[9] One sees the modernist temperament on display in contemporary American literature—two unforgettable characters come immediately to mind: Jay Gatsby and George F. Babbitt—and in popular culture from the 1920s onward. Ancestral wisdom, a reverence for the past, and skepticism about change in any form: the American experience largely repudiated these habits of mind, and in its stead embraced an orgiastic conflation of the future, with progress coupled with an eagerness to equate what is old with error and obsolescence. The idolatry of commerce could not help but influence the armed forces. Witness the founding of the Industrial War College in 1924, which was intended to offer courses on business administration, the large-scale purchasing of military supplies, industrial technology, and the like—in other words, fields of knowledge that would bear on mobilization and the provisioning of forces in the field. Strategy as an idea did not figure largely.

In the 1930s the War Plans Division—a legacy of Elihu Root—working in concert with the Army War College, produced the Rainbow plans that shaped successful operations in World War II. But even this brief and rather dim flickering of intellectual vitality soon disappeared beneath the exigency of how the war was fought. At the end of the day, we managed to defeat Imperial Japan and Nazi Germany as quickly as we did, because the war was determined less by maneuver than by attrition. Operational plans centered on logistics. In the Far East the chief challenge was not the possibility of a major counteroffensive by the Japanese, but our ability to deliver sufficient quantities of materiél at decisive points, while denying the same to the enemy. In Europe the situation was much the same. Victory hinged on the expeditious movement of fuel, munitions, and replacements of men and heavy goods to the front. By the time of Operation *Overlord*, the German armed forces had been greatly weakened by three years of hard fighting against the Soviet Union. Many of the best officers and NCOs, as well as the bulk of their equipment, had perished in Russia. It's worth recalling also that the German army's greatest defeat—far more catastrophic even than Stalingrad—was inflicted by a series of Soviet offensives during the summer of 1944. Between July and September, the *Wehrmacht* had absorbed more than two million casualties at a point in the war when they could no longer be replaced. Hardly less important is that by mid-1944, the political climate in Germany inhibited the operational decisions of the few remaining field commanders of proven ability. The coerced suicide of General Erwin Rommel in October of that year adumbrated the miserable state of the *Wehrmacht*'s battlefield leadership. The most efficacious way to win against the Germans was to overpower them in much the same way as a barrel-chested street brawler might pummel a bleeding and exhausted professional boxer who stepped from the ring into the back alley. To note this, of course, is not to disparage the heroism of long-serving volunteers and that of the citizen soldiers who fought beside them; rather it acknowledges the expedient approach—if often costly in lives and matériel—of fighting the war in a way that played to our strengths, while minimizing our weaknesses.[10]

One offshoot from World War II was the enshrinement of industrial processes in all aspects of military operations. The limits of this approach were demonstrated in the Vietnam conflict, but perhaps that lesson remained obscure on account of strategic calculation in the nuclear age—with its emphasis on the enormous destructive power of weapons that could be launched with the push of a button. What good are moral and psychological factors in a scenario shaped by mutually assured destruction? Given our historical experiences, it is hardly surprising that our planning method up to the present day betrays an indifference to things that cannot be measured, weighed, or reduced to a bullet statement.

An enormous volume of ink has been spilled, and a great many lungs exercised, in pointing out the breathtaking reforms carried out by the U. S. armed forces between the fall of Saigon and Desert Storm. There is much truth to this point of view, needless to say, but when it comes to the foundations of our war planning, not much has changed since the Second World War. Operational and strategic thought are rarely, if ever, in touch with each other. We continue to look upon war as a sports contest: you win or lose, grant interviews to the press immediately afterward, and then go home—leaving it to the sports writers and pundits to assign grander meaning to the outcome. Military officers in particular do not recognize that the summit of their profession has two peaks. One we never lose sight of: liquidating the enemy on the battlefield, or intimidating him with force so that he bends to our will without the necessity of combat. But the other peak is of equal importance: providing strategic advice to civilian authority. Military officers ought to begin operational planning with a clear-sighted grasp of the peace we and our enemy are seeking, and then reconcile ways and means with the ends identified by civilian leadership. This point of view is anchored in an act of sympathy: knowing the mind of the enemy. But military professionals too often look upon strategic problems as an engineering exercise, or a challenge that can be tamed by bureaucratic adjustments.

An egregious example of the desiccated thought that shapes our operational planning is the *Capstone Concept for Joint Operations* (August 2005), which is anchored in the assumption that a potential enemy can best be understood as a system. The *Capstone Concept for Joint Operations (CCJO)* offers an elaborate and rather woolly definition of what a system is. At more than 800 words, the definition is four times as long as the *Mayflower Compact,* about twice as long as the *Bill of Rights,* and half again larger than the *Emancipation Proclamation.* Yet the profuse expenditure of language manages to accomplish nothing apart from plunging the reader into mental darkness. The careless writing betrays a problem that is far more intractable than poor editing. The *CCJO* is chock-full of terms—"framework," "elements," "processes," "success mechanism," and so on—that describe activities that are automatic or unwilled; they are concepts that come to us from the engineering vocations, which cannot account for the moral, cultural, and intellectual circumstances that often determine the fate of an army in war. The *CCJO* relies on mechanical means to understand the defiantly unmechanical mind of man.

The authors of the draft *CCJO* seem vaguely, if imperfectly, to recognize the futility in seeking to understand an enemy by viewing him as a machine:

Although the systems approach is helpful in understanding the complex nature of a given target, it cannot account for all variables. Most systems will confound detailed understanding; their elements and processes cannot be accurately mapped; much of their inner dynamics will remain opaque to comprehension. Systems will often exhibit unpredictable, surprising and uncontrollable behaviors.[11]

If this observation is true, then what is the point of viewing the enemy as a system in the first place? Writing of this kind is typical of what bureaucracies produce: slipshod thinking buried beneath heaps of pseudoscientific bafflegab—sludge bereft of even a trace of ore.

There is nothing uniquely ham-handed about the *CCJO*. The most recent edition of *Joint Publication 5-0: Joint Operation Planning* (*JP 5-0*; April 2006) is of much the same character. *JP 5-0* is important because, at more than 200 pages, it amounts to a rulebook that comprehends terms, processes, and regulations. Commanders are not allowed to deviate from it except in "exceptional circumstances." It should come as no surprise that the document makes for traumatizing reading—unless one happens to be untaught concerning military and diplomatic history, or the history of military strategy as an idea.

Take for instance the manner in which *JP 5-0* defines and illustrates the relation between effects, objectives, and tasks:

> **Objectives** *prescribe* friendly goals. **Effects** *describe* system behavior in the operational environment—**the *conditions* required to achieve an objective. Tasks** *direct* friendly action.

The narrative reads like an instruction sheet for a household appliance or an insecticide:

> **Understanding the behavior of systems in the operational environment supports the use of effects in planning.** In the example above, effects 17–19 represent conditions for achieving objective 6, which relates to the President's first objective. Effect 17 also is a statement about the behavior of Red's military system, while effects 18 and 19 relate to the behavior of other regional countries' military, law enforcement, political, and social systems. **The full set of desired effects would represent the necessary and sufficient conditions for achieving the strategic objective.**[12]

The "objective" referred to here—one of eight which in total comprise an astonishing twenty-four effects—is to "Set favorable conditions for future stability and security in the region." The effects are as follows:

> E-17: Red is incapable of significant aggression against regional countries

E-18: Regional countries expand their counterterrorism training and capabilities

E-19: Regional countries view U. S. intervention and results as beneficial

These are notional examples and have not been taken from an actual plan. Even so, they illustrate how military planners are expected to think, and how military planning is taught as part of the various Joint Professional Military Education programs.

For starters, "behavior" is a grossly inapt idea in any discussion of military strategy. *Webster's* second edition tells us that the term is meant to apply to inanimate objects: the "behavior of a ship; the behavior of a magnetic needle," or the response of tissue to stimulation: in other words, the predictable reflexes of an organism unimpeded by emotion, memory, and an appetite for unreason. In ordinary usage, behavior suggests the deportment or manners of an individual person, though social scientists have co-opted the term in describing the mechanistic actions of groups or individuals, suitably typecast.

That the attitudes of regional countries—allies and neutrals—can be adjusted by military means is nonsense of the rankest sort. For starters, how can we expect planning staffs to take reliable measurements of whether "regional countries" view U.S. actions as "beneficial"? The real danger in establishing this kind of criterion is that it makes way for the substitution of hard strategic calculation by wishful thinking. One can imagine the discussion among military planners: "We can't get the higher-ups to sign off on our plan to neutralize Red Force unless we show that military actions x, y, and z will make countries a, b, and c view us in a favorable light. On the other hand, how can these countries not see us in a favorable light after we knock out Red Force." The truth may likely be that one of the countries whose good opinion we seek will seize territory or pursue an aggressive foreign policy in the aftermath of our victory, a development that might well undermine our national interests. Put another way, we should never assume—as we too often do—that our prospective enemies, allies, and rivals will see diplomatic and strategic developments in the same way as we do.

The difficulties are scarcely less nettlesome on the home front. The government of the United States cannot be expected to garner overwhelming domestic support for any military action, short of throwing an invader back into the Atlantic or Pacific Ocean. Any U.S. military operation overseas will generate domestic public debate, aspects of which will be discerning, dignified, and civic-minded, as well as foolish and laden with acrimony, and envenomed further by parochial interests of one kind or another. Far from being a regrettable thing, the absence of a unified national opinion is a sign, or at the very least a by-product, of self-governing societies. It is a distinctive trait of dictators to insist that their voice is the only authentic expression of national will. In advising civilian authority, planners should never ignore the diplomatic and political impact of military strategy.

In reflecting on the failure of the Western Allies to defeat Germany in 1944, General George Patton had this to say:

> [O]ne does not plan and then try to make circumstances fit those plans. One tries to make plans fit the circumstances. I think the difference between success and failure in high command depends upon its ability, or lack of it, to do just that.[13]

JP-05 encourages military planners to take the opposite view. The particularity of circumstance is expected to conform naturally to the template we've come up with—one that bears little relation to the way wars have been fought across time.

Because of the imponderables and indeterminate ramifications that attend any strategic enterprise, a military plan should be built upon an understanding of how an enemy—who is beholden not only to reason, but also to bigotry, self-sacrifice, and a host of cultural shibboleths—might view the world. Armies should concern themselves with breaking an enemy's will—a good way to accomplish this is by killing off a large part of his army—so as to leave him with no other choice than doing what we want him to (e.g., evacuate territory, cease organized resistance, disband, abdicate his authority and yield it to us). We should not care a whit about the casting of an enemy's various attitudes; the focus group and poll-taking mania is more useful to the popular press. Rather, we should care only about what he does or refrains from doing—and the only way to accomplish this is by waging or threatening war.

The right question to ask when faced with the notional crisis set forth in *JP-05* is, first, can the strategic objective—eliminate the threat to our national interest from Red Force—be realized by military means? If so, how do we employ military force to get Red to do what we want? By all means we should take into account material factors—logistical issues, weather patterns that might obstruct overseas deployment, the geography of the battlefield, the availability of suitable ports and airfields, and the like—but these calculations are relatively easy to make. More demanding is the supple thinking about which *JP 5-0* is breezily indifferent. Who are the Red Force's military commanders? How and where were they educated and trained? Have they published anything on strategy or tactics? What are the potential sources of friction among Red Force military commanders, and between the military and civilian leadership? What of the quality of morale of the troops they command? Do sources of friction here mean that formations made up largely of conscripts will be corseted by fanatical units, indicating that weak morale will be counterbalanced by strong discipline? Perhaps the generals of Red State are cautious, the civilian leadership reckless; perhaps the arrival of U.S. forces might be enough to cause paralyzing unrest between them. How likely is this to happen? And if it doesn't, are we prepared to invade and are we ready to install a puppet regime?

If we are not prepared to establish a government to our liking, and are not willing to buttress it for many years, then the invasion (or even the threat of invasion) are all for nothing. Which brings us back to the original strategic question: what compelling strategic interest is in play and are we prepared to pay a suitably high price—in treasure, blood, national morale, and diplomatic contention—to protect that interest? These kinds of issues highlight the synthesis of strategic and operational planning, which is something our current *JP 5-0*, for all its commentary on the primacy of strategy, does not take into account.

Hardly less important is this: what qualifies an officer to advise senior civilian or military leaders on these difficult matters? Certainly not the criteria used by promotion boards (which tend to take an unfavorable view of time spent in formal education programs that don't have an explicitly practical end) and the arrangement of assignments, in which officers are moved around within their specialty. Our military education programs do nothing substantial in the way of preparing officers to take on this kind of work, which goes to the heart of the matter. Nor does the emphasis on technocratic skill and bureaucratic savvy in our modern military bode well. It would take a measure of effort for the conscientious planner to triumph over the planted axiom in *JP 5-0* that war amounts to a contest between technicians using various mechanical artifacts, in the manipulation of an inanimate object.

Any great commander from the past would find our approach to war planning inscrutable, not because the tools of war have changed, but because of our obstinate belief that words can transmogrify war from what it is—an enormously complex human activity—into a humdrum engineering exercise. Operational planning is no business for the amateur or the technician, and it is probably impossible even for the military genius acting without the advice of a well-tutored staff. We need officers with a lively understanding of their profession—which begins with a firm grasp of military and diplomatic history—and not just those who have a keen grasp of their inevitably narrow career paths, and who demonstrate an unseemly eagerness to equate skill at operational planning with a familiarity with the latest bureaucratic jargon.

The shortsightedness of our strategic understanding can be illustrated by evaluating its principles by the lights of the war between Russia and Finland. This campaign puts the lie to the idea that one can plan a military operation on the assumption that the broader strategic and diplomatic consequences are transparent, and that they will naturally fall in line with our own aims. In 1939 the strategic priorities of Soviet Russia were not unlike our own. Though Stalin believed that the chief threat to the state came from within (i.e., political heretics determined to stand athwart history and impede the triumph of communism), he also recognized the necessity of acting aggressively as a means of forestalling an attack by foreign powers. "Preemption" is the term we use today. Shortly before the start of World War II, Stalin signed a nonaggression pact with Hitler that placed eastern Poland, Finland, Latvia, Estonia, and Lithuania within the Soviet Union's sphere of influence. Given the ideological hostility between the two countries that simmered beneath the official comity, the Nazi-Soviet agreement came as a surprise to diplomats the world over. But the agreement signaled not friendship, but strategic maneuvering that served each country's interests in the short term. Hitler was free to direct his attention toward France once Poland was carved up. Meanwhile, Stalin was given time to build up his armies and extend the limit of Soviet-dominated territory, in preparation for the inevitable struggle between National Socialism and Communism.

In October 1939 Stalin coerced Latvia, Estonia, and Lithuania into permitting Red Army forces to be stationed on their soil, thus establishing a buffer along the Baltic Sea. Stalin's eyes then turned northwestward, toward Finland, where he sought an adjustment of borders that would protect the port city of Murmansk from attacks launched from Petsamo and the Rybachi Peninsula, and that would also place Leningrad outside of artillery range from Finnish territory. Finland refused to comply, asserting through diplomatic channels that her neutrality would be threatened by any military cooperation with the Soviet Union, and might even provoke a German invasion.

Because Finland enjoyed stable diplomatic relations with Germany—the former provided Hitler's military machine with strategic metals (chiefly nickel), and the latter had demonstrated no aggressive intentions in the Baltic—Helsinki's refusal to yield territory was seen by Stalin as a de facto abrogation of the Russo/Finnish nonaggression pact signed in 1932. War thus appeared to Stalin as the only strategically sensible option. Given Russian military might—in tanks it outnumbered and in aircraft it equaled the combined armed forces of the rest of the world—the invasion was expected to go smoothly. The Russian high command assumed that overwhelming firepower would carry the day, and that a critical mass of Finns would, out of ideological sympathy, rally to the invasion and cooperate in deposing the Finnish government. The modest military cost of forcing Finland to accede to Stalin's demands would be well worth the benefit, and such diplomatic unpleasantness as might crop up would easily be brushed aside. Russian occupation of Finnish territory would not be to Hitler's liking, but his attention was absorbed by his war against the Western Allies.

Russian troops crossed the frontier on 30 November 1939. Straightaway things went badly for the invaders. Thanks in large measure to Stalin's purge of the military in 1937, Russian operations were poorly conceived and clumsily executed. The Finns concentrated their forces astride the most obvious invasion route, the Karelian Isthmus between Lake Ladoga and Leningrad, where the Soviets ended up delivering the main blow. The Red Army was stopped there and along most of the other seven axes of approach. Russian columns moving at a snail's pace through the snowfields and forests were forced to confront ferociously resolute attacks by Finnish troops moving freely across their home territory. The heavily wooded terrain, sprinkled with fen and lakes, and destitute of good roads, stressed to the breaking point the Red Army's tenuous lines of communications. Aggravating matters were the overly optimistic expectations of Stalin and his military commanders, who believed that Finland's tiny and poorly equipped forces would come apart like a cheap watch. More importantly, Finnish patriotism had been steeply discounted. In the wake of Finland's declaration of independence from Russia in 1917, a civil war between "Reds" and "Whites" ensued, with the anticommunist Tsarist sympathizers triumphant. The Russians falsely assumed that their invasion would be perceived as an attempt to correct the outcome of that conflict. Finnish communists would fight alongside them, and the bulk of the population would welcome Stalin's troops as liberators. Instead, the Red Army, along with the political creed it served, was battered and humiliated.

Diplomatically, events took an equally unpredictable turn. The stubborn defense of the Finns generated sympathy throughout the West, the world by then having seen enough of this sort of bullying from Hitler. A handful of volunteers from the United States, Hungary, and the Scandinavian countries fought alongside the Finns, thus putting a domestic face on what was for these countries a campaign of no major strategic consequence. More significantly, the Soviet floundering demonstrated to the world the ineptitude of her armed forces, putting to rest whatever misgivings Hitler might have entertained about waging war on Russia. The French and the British began assembling an expeditionary force to aid the Finns. This was ostensibly an act of solidarity with their struggle for freedom, but the real reason was that the conflict gave the allies an opportunity to open a second front against

Hitler on the Baltic, thus cutting off shipments of iron ore between Scandinavian mines and the Ruhr Valley. The potential loss of his supply of strategic materials, in turn, provoked Hitler to attack Norway and Denmark in the spring of 1940.

A peace treaty giving Stalin what he had originally sought by military means brought the war to an end on 15 March 1940. Even so, what had seemed to Stalin to be a necessary but undemanding campaign provoked circumstances that might easily have brought an end to the Soviet empire. In effect, Stalin encouraged Hitler to attack Russia and he nearly ended up making enemies of the very nations whose aid kept the Soviet Union from collapsing in 1941–1942. Had Finland managed to hold out for another few months, the Russians might have found themselves fighting French and British troops, and perhaps an increasing number of American volunteers, with consequences that one can hardly predict. The Russo-Finnish conflict is well worth studying for a number of reasons, not least of which is that it can teach us, derivatively, that our approach to campaign planning is misguided if it assumes that waging war is akin to a business or engineering enterprise.[14]

Evidence that our approach to war planning takes insufficient account of the complexity of war can be found in the way we prepared for Operation Iraqi Freedom (OIF), the invasion of Iraq. Combat operations were exemplary. Iraqi military formations were liquidated without any major setbacks, and the rapid envelopment and occupation of Baghdad, no less than the adaptability of soldiers in the field, has justly become military legend. The performance of the 1st Marine Division was particularly impressive. It is not insignificant to note that the division was commanded by Major General James Mattis, a man who embodies the intellect of Heinz Guderian, and the verve and intelligent audacity of Erwin Rommel. It was in the planning for postwar reconstruction of Iraq, where judiciousness was most needed, that the inadequacies of our current ways of doing business proved most conspicuous. Obviously enough, the Iraqi military would be no match for the armed forces of the United States. The difficulty would reside in replacing a dictatorship that had successfully imposed order on a collection of mutually antagonistic tribes. Otherwise, the expulsion of Saddam Hussein would be counterproductive. Military operations, in other words, were not anchored to broader strategic aims.

On this point in particular one sees the indivisibility of strategic and operational thinking and, correspondingly, our reflexive and apparently unwitting habit of drawing sharp divisions between the two. "[P]olicy cannot be separated from strategy," Helmuth von Moltke observed in his voluminous writings on war, "for politics uses war to attain its objectives and has a decisive influence on war's beginning and end." We would do well to contemplate Moltke's observation, but unfortunately our approach is quite the opposite. Read our doctrine, and the associated literature on military transformation, and one will get the idea that waging war is pretty much the story of technicians serving weaponry and computer systems. So far we haven't paid a heavy price for this, but sooner or later we will confront an enemy who is every bit as resourceful and powerful as he is barbaric—an adversary who has studied the culture of the United States as well as our military traditions, and will not be put to all that much trouble in discovering that our misbegotten belief that war is akin to an engineering problem is a weakness easily exploited.[15]

Senior officers often note that junior and midgrade officers don't think strategically, and perhaps this has to do with, among other things, a false understanding of the apolitical nature of military service. Active duty officers are discouraged from commenting on policy, for the time-honored and compelling reason that service members must not be seen as challenging the supremacy of civilian leadership on such matters. Even so, military officers are duty-bound to recommend to political authorities the best means of employing military force in the service of national, strategic aims. What follows is not meant as a criticism of contemporary political arrangements, but quite the opposite. The discussion is intended to exemplify how military professionals might think about the relation of operational warfare to grand strategy, and so effectively inform senior commanders, who in turn consult with the civilian leadership, on reconciling the application of armed force with national strategic objectives.

The state of Iraq, no different from much of Africa and other parts of the Middle East, was a creation of European cartographers, a circumstance that Saddam Hussein exploited for his own ends. Tribal discontents; the pan-Arabic nationalism (every bit as energetic as it was ill-conceived) that had sprouted in the wake of the Ottoman Empire's collapse; the Islamic world's instinctual sympathy for despotism, along with the mechanisms of tyranny perfected by European dictators: these things Saddam Hussein manipulated in order to achieve and consolidate power. Borrowing heavily from the example of Stalin, Hussein believed that the most imposing menace to his regime was not the United States or any other external power, but internal dissent. His security arrangements were designed to concentrate authority in his hands and to keep even his closest advisers largely ignorant of his political and strategic aims. Absolute loyalty was rewarded; merit and disinterestedness among subordinates aroused in Saddam Hussein disquiet, which, if conspicuously exercised, might well end in a death sentence or torture and mutilation followed by imprisonment. The effect of this strain of totalitarianism on governance is that legal, civic, military, and technical institutions would prove to be inept once left to their own devices.

The aimlessness and obtuseness of our strategy in Iraq remains an astonishing thing. To treat postwar planning as we did, with hardly a gesture toward the range of likely outcomes, reflects a willful ignorance of the ultimate aim of strategy, in which armed force is used to secure a better peace. We took no notice of circumstance, and instead justified the risks of using the smallest possible military force by assuming that Iraqi national and municipal organizations possessed a competence and vitality that was not validated by reason or experience. Once Saddam Hussein was eliminated, the most imposing (if not the only) challenge would center on rebuilding the physical infrastructure of Iraq, rehabilitating governmental services, nurturing a market economy, and rounding up isolated Baathist loyalists whose belligerence would be no more than a nuisance—the impotent spasms of a hated regime that had been decapitated. We rightly understood that Saddam was deeply unpopular, but we failed to come to terms with long-simmering tribal animosities among the Iraqis themselves.

It is difficult to refute the assertion that OIF was an imperfectly conceived operation. The details are set forth in *Cobra II: The Inside Story of the Invasion and Occupation of Iraq*, by Michael R. Gordon and General Bernard E. Trainor. A great volume of critical and historical literature has sprung up about OIF and its aftermath, much of it of perishable or doubtful value, but *Cobra II*, along with John Keegan's *The Iraq War*, makes for indispensable reading. *Cobra II* is first-rate history: clearheaded in its conclusions, artfully synthetic in the disposition of evidence, well written, painstakingly researched, and admirably free of academic and military jargon.

Gordon and Trainor claim that the secretary of defense asserted, indeed demanded, that operational planning must reflect (and thus the war on Iraq would validate) current ideas on military transformation. The view advocated by the secretary was very much in keeping with the American way of war and American culture in general: war is no different from any other physical phenomenon, and thus can be explained and disciplined by placing technology in the service of the latest theories. Our experiences, moreover, are unique—"we are fighting a new kind of war"—and so our means of confronting problems are, or cannot help but be, groundbreaking. Achieving America's strategic aims need no longer hinge on large numbers of well-supplied troops organized in conventional formations. Technology has eliminated much of the unpleasantness and previously imponderable aspects of ordinary life; why then can it not reform the messy business of war? In the future, so the transformation theorists would have us believe, decisive victory would spring from the expanded employment of Special

Operations Forces and, of much greater importance, technological wizardry in all its forms (weapons design, advances in intelligence-gathering apparatus, command and control innovations). Conducting warfare in this way would prove to be a triumph of efficiency (the progeny of modern scientism) over obsolete practices inherited from the industrial age. A military build-up in preparation for operations conceived along traditional lines would absorb a great deal of time, and require bothersome and uncertain diplomatic wrangling to secure access to ports, assembly areas, airfields, and the like—impediments that cannot help but inhibit the use of military force.

The technocratic outlook also squared neatly with the desire to depart from the path along which previous presidential administrations had moved deliberately, enthusiastically, and not without a measure of success. "In some nation-building exercises well-intentioned foreigners arrive on the scene, look at the problems and say let's fix it," Secretary Rumsfeld declared in a speech given in New York City the month before the war began. "This is well motivated to be sure, but it can really be a disservice in some instances because when foreigners come in with international solutions to local problems, if not very careful they can create a dependency." The operational planning for Iraq was expected to realize the following without incurring costs that can only attend a venture of this kind:

> The United States could oust a dictator, usher in a new era in Iraq, shift the balance of power in the Middle East in the United States' favor, all without America's committing itself to the lengthy, costly, and arduous peacekeeping and nation-building, which the Clinton Administration had undertaken in Bosnia and Kosovo. The new policy would be best for both sides, Americans and Iraqis, or so the theory went.[16]

This rather quixotic view was endorsed by "Operational Availability," a classified study conducted by the Joint Staff in the summer of 2002. The results of the study, which as Gordon and Trainor point out did not consider postcombat circumstances, seemed to provide transformation advocates with an up-to-date, scientific justification for waging war on Iraq with a relatively small force.

Others in a position to express an informed opinion were not so sure. General Tony Zinni had crafted the plan dismissed by Secretary Rumsfeld. General Zinni assigned a high priority to postwar issues, and in doing so assumed that a minimum force of 400,000 troops was needed to depose Saddam Hussein and effectively deal with the aftermath. According to the Zinni plan, the occupation of Iraq was expected to last up to ten years. In the months before the Second Iraq War began, a number of well-regarded voices echoed or concurred with General Zinni's conclusions: Secretary of State Colin Powell, Chief of

Staff of the Army General Eric Shinseki, former Assistant Defense Secretary and Navy Secretary James Webb, Steve Hawkins (an Army Corps of Engineers brigadier general detailed to the Joint Staff to help with stabilization planning), the RAND Corporation, and Robert Perito, an authority on stabilization operations who had extensive experience in the Balkans.[17]

Even the U.S. Army's official historian of the campaign, Major Isaiah Wilson III, came to much the same conclusion. In a study completed after major combat operations ceased, Major Wilson criticized not only the absence of a well-conceived plan, but also the Army's refusal to learn from experience. The report was summarized by Tom Ricks in the *Washington Post* more than eighteen months after the postcombat stage began:

> Wilson's essay amounts to an indictment of the education and performance of senior U.S. officials involved in the war. "U.S. war planners, practitioners and the civilian leadership conceived of the war far too narrowly" and tended to think of operations after the invasion "as someone else's mission," he says. In fact, Wilson says, those later operations were critical because they were needed to win the war rather than just decapitate Saddam Hussein's government.[18]

Major Wilson is hardly an isolated voice among the rank and file. A report by Tom Shanker and Eric Schmitt, published in the 23 April 06 edition of *The New York Times*, provides additional evidence for the argument that the Second Iraq War was poorly planned and that we seem not to have taken the right lessons from our improvidence.

What of the United States Central Command (CENTCOM) commander who would be tasked with realizing the secretary of defense's vision for twenty-first century warfare, in concert with arranging the expulsion of Saddam Hussein? General Franks had helped General Zinni in writing the plan that was rejected by Secretary Rumsfeld. Franks' efficiency and ardor prompted Zinni to recommend him as his successor at CENTCOM. But when the time came to plan a decisive victory over Iraq, Franks did not seriously oppose the views held by the secretary of defense, if only because he had no interest beyond moving troops to the theater of operations and winning the fight. "For Franks, anything other than war-fighting was an unglamorous and thankless burden," Gordon and Trainor observe. As combat operations wound down, General Franks "was already considering retiring and negotiating a multimillion-dollar deal to write his memoirs"; the stabilization of Iraq would thus become "somebody else's problem." At the end of the day, Franks was unwilling to express forthrightly, either in words or by resigning, what was at heart a philosophical difference between his boss and a critical mass of military opinion (which included his own, if we take General Zinni at his word), over how to reconcile ambitious strategic ends with inappropriate ways and transparently inadequate means.[19]

General Franks's manner of proceeding illustrates the weaknesses in how we have come to interpret the obligations of high command, specifically the estrangement of generalship from statesmanship. The cult of the manager and the technocrat has superseded the duty—sanctified by tradition and validated by Carl von Clausewitz and others—to act on the understanding that war embodies policy. The emphasis on technological advancement, especially in regard to command and control, likely will aggravate matters. The "Blue Force Tracker," a computer/satellite system that provides a digital representation of individual vehicles on the battlefield, is an alluring diversion for senior commanders (its many benefits notwithstanding), because it tends to draw them into an arena they know well, which is the Clausewitzian understanding of tactics: *"the use of armed forces in the engagement."* Gordon and Trainor point out that Franks often yielded to the temptation to run the war based on impressions taken from the Blue Force Tracker, even as he despised the more urgent, and what should have been for him the predominant, obligation to address strategic problems. Strategy, as Clausewitz points out, is *"the use of engagements for the object of war."* Franks's nonchalance (perhaps studied apathy is a more fitting description) toward the stabilization phase, and his unrelenting focus on combat operations (even though there was no chance that we would lose on the battlefield) reflects a dangerous and ultimately self-injurious frame of mind. The right question that Franks apparently refused to ask of himself and his staff was not how best to eliminate the Iraqi army, but what must be done to bring about the strategic end? A question all the more important to Franks, given that two months before the war started, President Bush had signed National Security Presidential Directive 24. This document gave the Department of Defense legal authority over stabilization operations, something Secretary Rumsfeld had been pursuing for some time.[20]

One sees the outlook of the managerial class and the technician on display in General Franks's memoir. *American Soldier* is indispensable reading for the historian, and certainly the nonspecialist reader will find it illuminating and quite often eloquent and affecting. Indeed, the reader is left with the impression that the officer directing the invasion of Iraq was blessed with an affability and reflexive kindliness that is rare among those who wield authority in a somber business. Even so, his recollections show us that modern generalship no longer owes a debt to the wisdom of Clausewitz. For starters, very little is said in *American Soldier* about the governing purpose of the campaign, which was not combat operations—a straightforward matter of applying suitable force against a feeble and disorganized opponent—but the replacement of one regime with another. Franks states that we might have done a better job planning for the aftermath of maneuver warfare, but he refrains from offering specifics. His silence on this point betrays a dislocated perspective on the great importance of the operational planner's responsibilities, rightly understood. "If we had it all to do over again—armed with what we know today—I'm sure some of the decisions would be different," Franks declares. "I am not at all sure, however, that all the decisions would be better." It is hard to interpret this remark as expressing anything other than a self-satisfaction, which, given the circumstances, is as puzzling as it is disheartening.

Elsewhere, Franks acknowledges the strategic necessity of the global war on terror, even as he notes the fecklessness of relying on gadgetry to achieve a decisive victory. "Today, simply because we fight with JDAMS, UAVs, the Blue Force Tracker, and satellite communications, does not mean that the primitive human emotions of pride, greed, jealousy, and xenophobic hatred have been extinguished," Franks observes. "Those negative human characteristics are unfortunately widespread in both Iraq and Afghanistan." What Franks says here is of course true, but one is left wondering why such thinking did not serve as a point of departure for the operational plan. "CENTCOM had many capabilities—engineering skills and equipment, medical teams, and Arabic-speaking civil affairs specialists," Franks notes. "But we had neither the money nor a comprehensive set of policy decisions that would provide for every aspect of reconstruction, civic action, and governance." Why, then, was not a compelling case made for these resources if such deficiencies would frustrate the objective behind going to war in the first place?

If Franks' memoir is any indication, there seems to be in play a fatalism of sorts in regard to how military officers view the shortcomings of Operation Iraqi Freedom. As Franks says, "history reveals that wars often end in chaos that continues for years after the last tank shell or artillery projectile is fired." Franks would thus have us believe that things have gone badly in Iraq postcombat operations, not because of inept planning, but because of ineluctable forces of history. Such a viewpoint is nonsense. It is the planner's job to confront chaos and either bend it to suit his ends or limit its impact. In Iraq we very well may have accomplished the opposite. Our chimerical strategic assumptions have created a mess that will take years to clear up, and, correspondingly, have simplified matters for our enemies in Iraq and elsewhere, as we have no choice but to continue expending political capital and military resources, until Iraq sits firmly in the camp of our allies.[21]

Supplementing *Cobra II* is Michael O'Hanlon's commentary on the planning of OIF. In *Iraq Without a Plan,* O'Hanlon notes that the civilian leadership was overly sanguine about the likely brutality and stubbornness of the chaos that an Iraq freed from the yoke of Saddam Hussein would entail, so ultimate responsibility for failure or miscalculation must be assigned to them. But O'Hanlon is equally critical of General Tommy Franks and members of the Joint Staff, who "gave their professional imprimatur to a military strategy that was innovative and solid for the invasion phase of the war," O'Hanlon declares, "yet negligently incomplete for the aftermath." Whereas Gordon and Trainor emphasize the culpability of the secretary of defense—who, in their account, subtly but persistently coerced military leaders into seeing things his way, and ignored solid advice generated by midlevel planners—O'Hanlon is less indulgent toward active duty officers. The intellectual and moral feebleness of joint force commanders and their staffs, O'Hanlon asserts, is deserving of censure. It is wrong, O'Hanlon argues, to blame "the mistakes of one civilian leader of the Department of Defense, and one particular administration, for a debacle that was foreseeable and indeed foreseen by most experts in the field." O'Hanlon goes on:

> Under these circumstances, planners and high-ranking officers of the U.S. armed forces were not fulfilling their responsibilities to the Constitution or their own brave fighting men and women by quietly and subserviently deferring to civilian leadership. . . . The country's Constitution makes the president commander in chief and requires military leaders to follow his orders. It does not, however, require them to remain mute when poor plans are being prepared. Nor does it require them to remain in uniform when they are asked to undertake actions they know to be unwise or ill-planned.[22]

There is an essential justice in what O'Hanlon says here. But a complementary explanation is that the culture of our planning operations pretty much guaranteed that we would act unwisely.

O'Hanlon rightly points out that enlightened and closely argued views were not in ascendancy among military planners, even though, as Gordon and Trainor note, there were isolated examples of clearheaded assessments of what lay in store for the United States.[23] Even the Army War College—which should have been a source for innovative thinking—offered up insipid advice on the eve of the campaign. In February 2003 the Strategic Studies Institute (SSI), under the sponsorship of the Army War College, published a monograph by Conrad C. Crane and W. Andrew Terrill. *Reconstructing Iraq: Challenges and Missions for Military Forces in a Post-Conflict Scenario* retails itself as an "analysis," but in fact its fourscore pages amount to a disjointed, superficial, and unoriginal survey of historical circumstances appended to a "matrix," which is nothing more than a motley pastiche of commonplaces. One might gather the same information from an hour or so spent on the Internet with a glossary of doctrinal terms at hand. The monograph is adorned with endnotes—more than one hundred of them—thus giving it the appearance of hard research in the service of an original point of view. The information, however, is corralled, rather than marshaled and interpreted. Most of the references are to governmental publications, with a collection of newspaper articles and a few academic books thrown into the mix.

The heart of the monograph is a collection of tasks that "must be accomplished to create and sustain a viable state." Ridiculous though the assumption may be that civil society resembles a tinker toy, the urge—instinctual to the bureaucrat and the technician—to impose a factitious order on a refractory and murky reality goes further:

> The 135 tasks are organized into 21 categories, and rated as essential, critical, or important for the commander of coalition military forces. They are then projected across four phases of transition—Security, Stabilize, Build Institutions, and Handover/Redeploy—to reflect which governmental, nongovernmental, and international organizations will be involved in execution during each phase.

The authors go on to say that although their recommendations focus "specifically on Iraq, these insights [sic] will apply to any important post-conflict operation."[24] The argument of the monograph, as this passage demonstrates, is hobbled by simpleminded hubris. Never do the authors bother to consider the aspirations, abilities, and competing historical memories of the conglomeration of groups that, with a bit of engineering expertise (so the authors would have us believe), are expected to coalesce into a nation-state devoted to individual liberty, religious tolerance, and free markets. The authors assert that to succeed, all we need to do is bolt together the raw materials—the Iraqi people and bureaucratic machinery—by the lights of the matrix. It is worth comparing this treatise on postmilitary stability with our planning for the occupation of Germany. Earl F. Ziemke's book on the subject shows U.S. military and political leaders focusing on the possible behavior of the citizens of a defeated nation. Sixty years later, our concern is not with people, but with tasks.[25]

The intellectual aridity of our strategic calculations is on display in Afghanistan and Iraq. In March 2006 an Afghan citizen faced the death penalty for converting to Christianity, thus raising the possibility of a judicial murder committed by a political authority that owes it existence to American military and diplomatic power. On 23 March 2006 the U.S. secretary of state talked directly with the Afghan president about the issue. Shortly afterward the charges were dropped on technical grounds (supposedly the Christian convert was mentally unfit for trial), a decision that angered Afghani religious leaders. "Senior clerics in the Afghan capital have voiced strong support for the prosecution," a leading news service reported on 24 March 2006, "and have warned they would incite people to execute Rahman unless he reverted to Islam." Italy agreed to grant asylum to the 41-year-old man, but even this angered Afghani clerics who vociferously protested his departure. The Afghan constitution nominally guarantees religious liberty; whether that particular idea was inserted as a means of placating the sensibilities of the United States, or as a genuine attempt to break away from or at least temper Islamic legal tradition in which apostasy is a capital crime, is unclear. What we can say for certain is that such incidents are inevitable and will tax the goodwill of the United States, as well as undermine the authority of any Afghan government that seeks to maintain it.[26]

The situation in Iraq is far worse and poses a grave threat to American strategic aims, and to the political fortitude necessary to realize them. Despite the stalwart valor of American soldiers, Iraq remains a mess nearly four years after major combat operations ceased. Rarely a day goes by without the eruption of conscienceless mayhem. Iraqi security forces lack cohesion; desertion is rife. The democratically elected government, whose primary duty is to establish domestic order by exercising a monopoly on the legitimate use of force, appears feckless in its efforts to discharge this basic obligation. Pundits talk of the country being on the edge of civil war, but even this grim prospect seems optimistic. Civil war suggests agents seeking power by violence. But the taking of life seems a casual affair, or purely negative in its intent—the expression of feral cruelty and atavistic sadism masquerading as religious zeal or tribal solidarity.

Even if the situations in Iraq and Afghanistan improve dramatically, there is a deep-rooted problem with our strategic and operational planning method that must be addressed, or we will end up exhausting our treasury and our national will in pursuing objectives that can never be attained. In planning for regime change in Iraq, the armed forces blundered in ways that bring to mind Barbara Tuchman's criteria for political and military folly. First, the mistakes must have been apparent to contemporaries and not determined to have been so by the retrospective judgment of the historian. Second, efficacious alternatives must have been spurned or overlooked; folly must be recognized as springing from will and not circumstance. Third, the error cannot be attributed to one person alone or to a collection of advisers, but to a generational or cultural outlook.[27]

The debate surrounding the force levels needed to achieve decisive victory in Iraq illustrates the folly of our current approach to war planning, and is not merely a demonstration of the rarity of perfect foresight in strategic matters. About a month before the Second Iraq War commenced, Chief of Staff of the Army, General Eric Shinseki, testified before the U.S. Congress that nearly a half-million troops would be needed to pacify Iraq and begin the enormously difficult task of reconstituting a government sympathetic to American strategic interests. This was hardly an eccentric argument: General Zinni had reached the same conclusion many years earlier.

Even so, Shinseki's views were contrary to the outlook that obtained among senior Pentagon decision makers, in which overwhelming firepower, technical proficiency, and the allure of American political ideas were expected to carry all before them. There is some virtue in such a viewpoint so long as it is informed by the wisdom found in the works of esteemed authorities such as Carl von Clausewitz. Clausewitz does not discount the physical element in war, even though he rightly argues for the primacy of psychological and cultural factors. Superiority in numbers, "far from contributing everything, or even a substantial part, to victory, may actually be contributing very little, depending on circumstances," Clausewitz declares:

> But superiority varies in degree. It can be two to one, or three or four to one, and so on; it can obviously reach the point where it is overwhelming. In this sense superiority of numbers admittedly is the most important factor in the outcome of an engagement, so long as it is great enough to counterbalance all other contributing circumstances. It thus follows that as many troops as possible should be brought into the engagement at the decisive point. *Whether these forces prove adequate or not, we will at least have done everything in our power. This is the first principle of strategy* [emphasis added].[28]

Our strategy in OIF, the fruit of transformation theory, comes up short when measured against Clausewitz's reflections on the nature of war. For us, armed conflict has come to resemble a business transaction: profits are maximized when executives keep inventories lean. But war doesn't work this way, as military history makes plain. A force in being—air and naval fleets and concentrations of fresh troops—makes no sense to the business executive, but in war it can alter the strategic calculations of an enemy. Strategic and operational reserves must be ready at hand, even when large numbers of troops are not employed to force a decision in the quickest possible time. Reserves are used either to reinforce success or to eliminate or mitigate unforeseen setbacks. Our strategic decision during OIF of "off-loading" units, that is, aborting a unit enroute to a theater or canceling deployment orders in the name of efficiency, proved to be a mistake that might easily have been avoided by a disinterested consideration of sound military theory and history.

Tuchman's second criterion, that attractive options must have been rejected for a strategic decision to qualify as folly, was evident not only within the Department of Defense, as Gordon and Trainor point out, but also in the preoccupation with Iraq at the strategic and diplomatic level. That the subjugation of Baathist Iraq was an urgent and necessary task in winning the global war on terror was an attractive, but misbegotten assumption. Iran, with its rapidly maturing nuclear program and its sponsorship of terrorism in the Levant and elsewhere, stood as a more compelling threat to regional stability and American security. The same can said about North Korea. Encouraged by the benevolent indifference of China, Kim Il Sung's regime continues openly and defiantly to experiment with ballistic missile technology, something that has caused one diplomatic crisis after another over the past few years.

Even if one were to accept the idea that Saddam Hussein possessed weapons of mass destruction (WMD), he did not have the capability of directly attacking the United States. And surely the prospect of immediate and massive retaliation would have, or did, prompt him to have second thoughts—a calculation that is plainly beyond the intellect of the North Korean dictator, who stands as the most reckless head of state in recent times. Yes, the possibility exists that Saddam might have acquired WMD and used terrorist organizations as proxies to harm the United States or disfigure its national honor. But in 2003 we were already making progress in stopping terrorist attacks against the United States. That we have not suffered a single attack on U.S. soil in six years testifies to the success of our efforts at forestalling a repeat of 9/11. If military confrontations with Iran and North Korea were not to our liking, it would not have been imprudent to concentrate our limited resources—people, money, the support of the citizenry—on building our defenses at home, particularly systems that can thwart or forestall a nuclear attack, rather than to invade Iraq, knowing as we did that the rebuilding effort would be a dauntingly nettlesome task. Thus, we put aside something that was important and of immediate value, and instead pursued a course for which the payoff was remote and uncertain. To say this is not to argue that removing Saddam Hussein was utterly pointless; but there is no reason why it was not apparent to planners that reconstituting Iraq would be, prima facie, a treacherous and exhausting endeavor.

It is somewhat surprising that many observers correctly point out the differences between the post–World War II occupations and OIF (the military dictatorships of Japan and Germany enjoyed a measure of popular support that Saddam Hussein could never match), but in doing so they tend to ignore or downplay the similarities. Building a nation in the wake of a major war demands large numbers of troops, adequate—which is to say, profligate—funding, and an equally generous reservoir of national will and patience. Like postwar Japan and Germany, Iraq in April 2003 had been deprived of a government that, for all of its perfidy, carried out with ruthless and efficient zeal the first duty of political authority: the maintenance of order. The only way to reconstitute a culture, which is what nation building really means, is to employ overwhelming force. Victor Davis Hanson and other esteemed proponents of our current approach claim that a large number of troops are proportionally more vulnerable to terrorism. This is of course true, but casualties are one of those nasty realities of war, and in the long run the employment of a suitably large force would likely have proved to be less costly than our current strategy.

The half-million troops that General Shinseki and others advocated would have been more effective at policing the country, liquidating the enemy, sealing the borders, and (scarcely of less importance) avoiding the great evil of facing the unhappy choice of a gradual military build-up or, worse yet, a strategic withdrawal forced upon commanders by domestic politics. The 2008 presidential election may install an administration that finds political and moral satisfaction in terminating our involvement in Iraq. Much political hay has been made over the danger of establishing a timetable for the departure of military forces from Iraq, but the terrorists in Iraq know that the U.S. election cycle works in their favor. All they have to do is allow to American forces Schlieffen's "ordinary victories," in which the terrorists absorb casualties and make no appreciable advances, but avoid an irreversible strategic defeat. The decisive moment will come when a new administration enters office. No American president will want to be seen as escalating our involvement in a theater that is increasingly seen as remote from America's vital national interests. However charismatic or skillful the current or subsequent administrations may prove to be, the weight of America's culture will tell. Americans come to war reluctantly; they prefer conflicts to be short and they expect fighting to yield clear, demonstrable benefits—as in the case of the war against Japan and Germany and the postconflict reconstruction of those countries.

In World War II we demanded unconditional surrender from Nazi Germany and Imperial Japan, and in so doing we declared war on their national identity. A critical mass of both countries' male population was either killed or maimed in combat, thus setting the stage for a complete rehabilitation of their political cultures. We committed vast resources to rebuild Japan and Germany, because we had learned the consequences of indifference and vindictiveness from post–World War I Europe. Soviet imperialism, moreover, threatened both countries, and our own survival would have been jeopardized had we not intervened in an aggressive way. For many years afterward we took responsibility for the self-defense of Germany and Japan and, in effect, their economies and diplomacy as well. We participated in legal proceedings that convicted and punished wartime political and military leaders, applying the death penalty in many cases. We imposed a (very good) political constitution on Japan. And it's worth remembering that General MacArthur authorized activity that today would generate scandal. MacArthur's mission was "more profound and grandiose than political reform," writes Ian Buruma in his book, *Inventing Japan, 1853–1964*. "Japanese culture itself, the entire cluster of Japanese mentalities grown over thousands of years like noxious weeds, had to be overhauled, cleansed, and remade." School textbooks that contained references to Japan's military past were censored, songs expressing contempt for the occupying powers were banned, and so on.[29]

Advocates of our current involvement in Iraq have plausibly argued that the front line of the global war on terror (GWOT) is in Iraq, but its resonance with the American people remains uncertain. Any number of reasons might explain the public's growing apathy and pessimism about the OIF, including the steady dribble of casualties, the virulence of the unabated terrorist attacks directed at Iraqi civilians, and the disappointingly modest accomplishments of the U.S.-trained Iraqi security forces.

In comprehending the cast of American public opinion on this point, one need only consult our experiences during the Great War. President Wilson, expressing a view that obtained throughout the United States (New England excepted), refrained from waging war even after German U-Boats began torpedoing ships that carried the Stars and Stripes, as well as those that flew the banners of other neutral countries. Americans "had got used to maritime atrocities, had grown accustomed to official crises over ship sinkings," observes Barbara Tuchman:

> The *Lusitania*, the *Sussex*, the *Arabic* had followed one after another, provoking Wilson's notes, [Secretary of State William Jennings] Bryan's resignation, endless correspondence in incomprehensible diplomatic language, even some quite comprehensible threats and ultimatums, all mixed up with similar eruptions vis-á-vis the British contraband and blacklisting. It was all very confusing and—to the majority of the country—remote.

What generated a popular urge to fight was the famous telegram written by Arthur Zimmermann, the Kaiser's foreign secretary, and published on 1 March 1917, in which Germany sought to arrange an alliance with Japan and Mexico to wage war on American soil. The telegram, Tuchman writes:

> was not a theory or an issue but an unmistakable gesture that anyone could understand. It was the German boot planted upon the American border. To the mass of Americans, who cared little and thought less about Europe, it meant that if they fought they would be fighting to defend America, not merely to extract Europe from its self-made quarrels.[30]

The parallels with our own day are not hard to miss. The mass murder of American citizens on 11 September 2001 energized nationwide support for immediate military action. The photographs, not only of the burning Twin Towers, but also of the faces of the perpetrators, mostly Saudi Arabian nationals, are impossible to forget. But the public mind finds it increasingly difficult to connect those images with a war that carries not the name of a flesh-and-blood enemy, but of an abstraction, the "global war on terror" (a better term might have been, "The Twin Towers War"), the front line of which, the public is told, is in Iraq.

What, then, were the alternatives that were ignored or dismissed? If our strategy had given thorough consideration to reconciling ends with ways and means, we might have taken either of the following two options. We might have conquered Iraq and established an occupation force along the lines of those used in Germany and Japan after the Second World War. Accomplishing this would have required a rapid and large expansion of the armed forces—troops, matériel, and infrastructure—so that we could conduct operations in Iraq at a suitably intense level, while also maintaining a credible force to meet threats that might emerge elsewhere. If the GWOT is a fight for our survival, and Iraq is the front line in that war (as *The National Strategy for Victory in Iraq*, November 2005, claims), then deploying 500,000 troops should not have been peremptorily rejected as an extravagance. By way of comparison, on V E Day General Eisenhower had at his disposal nearly two million troops available for immediate occupation duty. At the cessation of hostilities these troops preserved order, and, more importantly, by their very presence demonstrated that the authority and influence of the Nazi party had been eradicated. As Earl F. Ziemke points out, the only major flaw in blanketing Germany with American soldiers was the "impermanence" of such a large force, as most of these troops were to be reassigned to the Pacific Theater or demobilized. On the other hand, if a large army turned out to be politically infeasible or not to the liking of senior policy makers, we might then have continued to keep Saddam Hussein hemmed in militarily, diplomatically, and economically. Neither was an appealing choice, but the path we chose to pursue—an attempt at nation building with an expeditionary force—is arguably the least attractive of the three options.[31]

Given our limited political and financial resources it would have been wiser to fight within the means available to us, instead of counting on the miraculous: that the obvious impediments to our objectives would melt away or never materialize. This is hardly an original argument. In an editorial piece that appeared in the *Washington Post* many months before the war started, former Assistant Secretary for Defense and Secretary of the Navy James Webb argued that a strategy of containment appeared the most efficient way to deal with Iraq. A unilateral war that does not take into account the years-long occupation that conquering a country demands cannot succeed, Webb suggests in his 4 September 2002 opinion piece. A "long-term occupation of Iraq would beyond doubt require an adjustment of force levels elsewhere, and could eventually diminish American influence in other parts of the world." Webb's concluding remarks warrant careful consideration by those who would argue that invading Iraq was a compelling necessity. "Unilateral wars designed to bring about regime change and a long-term occupation should be undertaken only when a nation's existence is clearly at stake."

And even if we decided that Saddam Hussein must be sent to a twenty-first century version of St. Helena, our recent experiences in the Balkans—General Shinseki had served as commander, NATO Stabilization Force in Bosnia-Herzegovina, during the late 1990s—should have given us at least an inkling of the obstacles facing us. We did not have the troop strength, or suitable numbers of the right kinds of troops (e.g., civil affairs specialists, linguists, and the like) to remake Iraq in a politically seasonable span of time. Nor did we have a clear idea as to how our involvement could bring about a workable government friendly to the United States. In a word, we failed to meet the fundamental obligation of military planning, which is to reconcile our strategic ends with appropriate ways and means.

In light of the constraints that circumstance imposed on us, a more effective way to address the terrorist threat would be to have followed Webb's advice, not by deposing Saddam, but by keeping him under wraps, and, in terms of fighting the GWOT, by stopping agents of destruction from permeating our borders and by reconciling our military strategy with suitable means. The limited resources spent on Iraqi reconstruction might have been better used in this way; and the political capital used to justify our ongoing involvement in Iraqi affairs might have been used to expand our armed forces, or at least reverse the imprudent decline in force levels, rehabilitate our gutless and inept immigration policy, and bolstered our intelligence operations at home and abroad.

One could argue against our current prosecution of the GWOT by asserting that the terrorists most likely to do us harm are not in Iraq, but currently residing in Dallas, Tacoma, Atlanta, Milwaukee, or Jersey City, furtively preparing the next strike. It is not unusual to read news accounts of gestating terrorist plots aborted by the diligence of the FBI. In March 2007, for example, the FBI issued a bulletin to municipal and state authorities warning that foreign extremist groups were seeking to acquire school bus driver's licenses.[32] Defenders of our current manner of proceeding might claim that creating a new agency—the Department of Homeland Security—suitably addressed the threat of *jihadists* operating within our own borders. But this kind of response—which is the first resort of that distinctively modern creature, the bureaucratic man—does little good and some harm. Aside from their great expense, bureaucracies slow things down by needlessly expanding the levels of policy review, by increasing the scope of infighting among government agencies competing for recognition and money, and by exponentially increasing the parochial self-seeking that is the enemy of efficiency.

In Iraq it wasn't long before the frailties in our strategic and operational planning became conspicuously apparent. We fought for "peace" and the "spread of liberty"; and though war by itself can accomplish much, it cannot give concrete form to abstractions. As Colonel Reynolds notes in his history of the Second Iraq War, American commanders emphasized to their troops that the war was not being fought on behalf of American self-interest, but to liberate the Iraqi people from tyranny. A great fuss was generated when television cameras broadcast to the world an American soldier placing a U.S. flag over the statue of Saddam Hussein in Firdos Square, Baghdad, on 9 April 2003. Within minutes the Stars and Stripes disappeared from view. "The flag that mattered was not the American flag," Colonel Reynolds observes, "it was the Iraqi flag, flown by itself, in sites no longer dominated by statues and murals of Saddam Hussein." Here we have a particularly dramatic expression of the intellectual friability of the entire enterprise. The display of the American flag in Firdos Square should have mattered a great deal.[33]

Because we defined the invasion of Iraq as vital to our national security, the American flag should have stayed in place along with a sizable occupation force, until a functioning Iraqi state sympathetic to the United States came into being. The political reasons for acting as we did are plain enough. We did not want the Iraqis, regional states, our allies, or the American people to interpret our victory as a conquest. But this is naïve over-scrupulousness at best; at worst, it is inept diplomacy. No disinterested observer would believe for one second that the United States nurtured a desire to establish a colony in the Middle East. The nation that has tolerated a dictatorship in Cuba for nearly two decades after the collapse of its patron, the Soviet Union, and that yielded sovereignty of the Panama Canal, could hardly be expected to covet Iraq. And no matter how strenuously we trumpeted our altruistic intentions, our rivals and enemies would seize the opportunity to protest what political expediency, rather than a fondness for liberty, impels them to call an example of America's naked aggression.

The flag flap prompts unfavorable comparisons between our current strategic planning and our way of facing down previous threats. Instead of supporting a proxy army that served our strategic interests—which was our preferred method during the Cold War—in Iraq our armed forces became a proxy for the belief, or rather the delusion, that the Iraqis were Americans at heart in their political attitudes, even though there was no convincing evidence for this. American troops fought for their country and their fellow soldiers; their lives were not, or should not have been, put at risk in the service of a chimera. Saddam Hussein was expelled because the Iraqi state, which he embodied, posed a threat to the United States. That the United States had an interest in postwar Iraq should not have gone beyond guaranteeing that the cure—the successor to the Baathist regime—was less dangerous than the disease. When U.S. Marines stormed Mount Suribachi (unlike previous islands taken by American forces, Iwo Jima was not an imperial possession, but a part of Japan), they planted the Stars and Stripes when they reached the top—not the Rising Sun or some item that represented freedom or peace in the abstract.

It is also worth noting that the Iraqi flag was not an ancient expression of a national identity lately suppressed by Saddam Hussein. Rather it was the very flag that he designed, or at least modified in 1991, the previous flag having been created in 1963 by Saddam's Baathist predecessors. So we deposed the man but protected the symbol of all that he stood for. That the display of the Stars and Stripes was seen as an impolitic act reflects the folly of wishing for an end—regime change—without supplying the appropriate ways and means to bring it about.

Tuchman's third criterion, that diplomatic and military folly cannot be laid at the feet of a single person, but must reflect time, place, and accumulated circumstance, brings us to our most disquieting observation: the principles that govern our approach to strategy are badly out of date and may ultimately do more harm than good, despite the altruism that heavily informs them. Our current National Security Strategy (NSS) is faulty in ways that can only enfeeble derivative policy statements (such as the National Defense Strategy and the National

Military Strategy), operational planning, and—as we have seen in Iraq—actual fighting.

The chief weakness of the National Security Strategy is that it is not a strategy at all. For starters, the NSS tips the scales at fifty-four pages, which is akin to a graduate-school thesis, rather than a cardinal statement of a nation's diplomatic position and political outlook. A national security strategy, rightly taking into account the primacy of circumstance and the necessity of balancing risk, should be expressed in as few words as possible. The language should be abstract, yet unambiguous—thus allowing political and military authorities a great deal of latitude in regard to handling crises. Perhaps the following is not possible given the bureaucratic dogma in which prolixity equals substance, but a wholly satisfactory national security strategy might be conveyed by a couple of sentences. The government of the United States is obliged to protect the lives and property of its citizenry, a duty that no international agreement or convention shall be allowed to supplant or corrupt. The United States will defend its interests whenever they are threatened. Circumstances will determine whether force or diplomacy, or a combination of both, is used to settle matters.

It scarcely needs mentioning that this definition does violence to the American political sensibility, which sees diplomacy—and its natural appendage, military force—as an expression either of America's missionary zeal (politically we are the world's exemplar, and thus are obliged to impose our enlightened institutions indiscriminately across the world), or her isolationist tendencies born of our historical experience (the old world is irredeemably corrupt, so we are better off keeping it at arm's length; therefore, our diplomatic and military efforts must be essentially defensive). In weighing the validity of this point of view, we ought to consider the circumstances of the Great War. The United States declared war on the Central Powers, according to President Woodrow Wilson, in order to make the world "safe for democracy"—a coinage that has stood the test of time and at some level is viewed as a justification for our current strategy. But then and now this is a falsehood easily uncovered. Our enemy, Germany, was at least as democratic as Great Britain and France, while one of our other allies, Russia, was not. And the Treaty of Versailles all but guaranteed that democracy in Germany would provoke circumstances that the victors would find themselves ill-prepared to meet.

That a strategy based on dispassionate self-interest departs from the Wilsonian tradition of American diplomacy does not make it ineffectual or reprehensible; in fact, it is perfectly in keeping with the Monroe Doctrine, which was essentially abandoned when the United States became a major player in world affairs in the aftermath of the Great War. "It is only when our rights are invaded or seriously menaced that we resent injuries or make preparation for our defense." Thus spoke President Monroe to Congress, 2 December 1823. In communities that share a common culture, morality has its place as a foundation and supplement to law; otherwise, morality can only be the concern of individual men and women. Self-interest backed by force is the ultimate governor of international politics. Indeed, there is nothing revolutionary or even iconoclastic about such a strategy (history shows us that abstract considerations of right and wrong have no place in the world of international politics) and the times we live in argue compellingly for its propriety. Apologists for the current NSS might assert that our unique experiences justify the lofty moralizing that saturates and fuels the document, but the effect may very well be quite the contrary. We may end up spreading ourselves too thinly, or we may refrain from taking action that is, by the light of our moralistic outlook, unsavory, even though our national interest demands it. More likely, we will end up confounding ourselves and frustrating our allies.[34]

History is a useful guide, but we seem not to have consulted it as thoroughly as we should have done. In the past the sort of stability that the United States professes to seek has been accomplished by one of two means: by establishing an empire, through conquest, rule, and exploitation, or through balance-of-power diplomacy in which expansionist or aggressive tendencies are kept in check by countervailing alliances. Clearly, empire is neither desirable nor practicable. The time has come for us to pursue a national security strategy based on a balance-of-power calculus, rather than the gratifying, but unworkable, commitment to noble abstractions that take far too little account of how the world actually works.

We remain the world's greatest military power, but we no longer exert the same degree of influence that we enjoyed in the decades after the Second World War, when our unprecedented economic vigor and the menace of Soviet expansionism compelled or encouraged many nations to court our favor. In his discussion of the historical context of "preemption," John Lewis Gaddis makes the point that much of our diplomatic influence during the Cold War derived from the prospect of "something worse." Nowadays, that prospect has receded from the view of much of the world, and our national security strategy should take fuller account of that fact.[35]

In his recent book, *The Battle for Peace*, General Tony Zinni (USMC, ret.) offers a sensible definition of what strategy is: "how a nation chooses to engage with the world." Zinni's point of view cannot be lightly disregarded. A highly decorated veteran of nearly four decades of military service, General Zinni's valedictory assignment (1997–2000) placed him in charge of the U.S. Central Command, the responsibility of which comprises more than two-dozen countries in the Middle East and nearby areas. General Zinni and his staff created the operational plan for dealing with Iraq that his former subordinate and successor, General Tommy Franks, chose to cast off. General Zinni argues that there are two ways to approach strategy: we can decide what we hope to see in the world as an ideal, and try to make it conform to our vision. Or, "we can take a hard, cold look at the environment to see how we can shape our strategy to achieve the best possible goals within the limits of the environment and of our power and influence." The pivotal term here is "possible," which implicitly admits to the enormous obstacles that are always in play. Our current approach bears more of a resemblance to the former, rather than the latter, of Zinni's sensible ideas on strategy.[36]

The 2006 NSS reads like a blend of an investment firm prospectus and sectarian covenant, rather than a clearheaded statement of strategic objectives. Witness the first sentence: "It is the policy of the United States to seek and support democratic movements and institutions in every nation and culture, with the ultimate goal of ending tyranny in our world." It is unreasonable to assume that one nation might accomplish a goal of this magnitude. How, for example, are diplomats—let alone military planners—expected to realize this end with the means available to them? And what manner of proceeding might square with ends that amount to the recasting of human nature? The "essential tasks" around which the NSS is built hardly help matters:

- Champion aspirations for human dignity
- Strengthen alliances to defeat global terrorism and work to prevent attacks against us and our friends
- Work with others to defuse regional conflicts
- Prevent our enemies from threatening us, our allies, and our friends with weapons of mass destruction (WMD)
- Ignite a new era of global economic growth through free markets and free trade
- Expand the circle of development by opening societies and building the infrastructure of democracy
- Develop agendas for cooperative action with other main centers of global power
- Transform America's national security institutions to meet the challenges and opportunities of the twenty-first century
- Engage the opportunities and confront the challenges of globalization

A few of these tasks are highly ambitious though not beyond the compass of reason, but several items on this list can aptly be described as fanciful. The most controversial section of the NSS, "The Way

Ahead," asserts that freedom (a murky term, the worthiness of which depends upon the context) is a fitting and proper strategic objective, because such a policy "reflects our values because we believe the desire for freedom lives in every human heart and the imperative of human dignity transcends all nations and cultures."[37] Such a point of view is bereft of curiosity and humility.

Elsewhere the NSS makes use of the term, "transformational diplomacy," which appears to be either an oxymoron—diplomacy seeks advantage incrementally when it is not devoted to guarding the status quo—or a euphemism for imperial pretension:

> Transformational diplomacy means working with our many international partners to build and sustain democratic, well-governed states that will respond to the needs of their citizens and conduct themselves responsibly in the international system. Long-term development must include encouraging governments to make wise choices and assisting them in implementing those choices. We will encourage and reward good behavior rather than reinforce negative behavior.[38]

What might readers abroad make of this passage? The argument here sails dangerously close to a metaphor that damages our ability to exert influence at times and places of our choosing, with the United States as an over-scrupulous parent, eager to encourage her children because disciplining them is painful to contemplate, and betting that her manifest concern will restrain their native waywardness. Facetiousness aside, it is clear that a legalistic frame of mind informs the document and assumes that contention among nation-states springs from misunderstandings, rather than the clash of national interests. Note the language: "our partners;" international relations characterized as a "system," etc., whereas elsewhere in the NSS, chaos, aggression, and violent bigotry are referred to as "security and stability challenges." But international relations are not, and can never be, governed by law. There is no such thing as an "international community." Rather, the world is comprised of a collection of disparate powers acting without regard to anything other than self-interest, fear, resentment, and prestige. Given that we are no longer the colossus that we were fifty years ago, it would be prudent if our strategy were to recognize the flinty truth beneath the rhetorical frippery.[39]

All things considered, the current NSS reflects too narrow a perspective on contemporary geopolitics, something that can be a source of anxiety for our allies, and that might also inject paralyzing confusion into military planning. Enemies might see a crusading spirit on display here that can be repackaged for their own domestic political ends, such as, for example, that our stance embodies imperialism and poses a threat to time-honored folkways and nationalistic self-determination.

Thus by its own pronouncements the United States becomes a threatening presence, which any foreign demagogue can blame for domestic or regional contention. Allies, acknowledging the odds that their populations would distrust close ties with a domineering America, might weigh the possible consequences of a great power that sets too high a standard for its ability to shape world affairs, and so is vulnerable to demoralization in the wake of strategic overreaching.

A plausible objection to balance-of-power calculations—which takes for granted the primacy of nation-states in international affairs—is that it takes no serious account of the terrorist threat. But this is not quite true. However vicious they are, and no matter how much they are beholden to unreason and tartarean violence, *Al Qaeda* and their ilk cannot harm us without the help of nation-states such as Iran—which cannot possibly benefit from getting into a fight with the United States. Much has been written about the decline of the nation-state and the corresponding rise of "nonstate actors," but this seems to be an overly ambitious proposition. The political necessity of statehood is obvious: it is the only way by which man can achieve and wield power, accumulate wealth, and attain a measure of security. Nonstate actors can never be more than clients or proxies of established states.

We also should work to avoid unwittingly cooperating with our enemies by making ourselves vulnerable. It's worth recalling that the 9/11 terrorists learned to fly in the United States. Many of them were educated and resided either here or in Western Europe. We have in part rectified the circumstances that allowed them to live among us, but more needs to be done in reforming our immigration policy. A good start would be to liberate our screening procedures at ports of entry from ill-conceived racial sensitivities and instead concentrate more intelligently on detecting the terrorists themselves—a lesson that has long been validated by Israeli practices.[40]

Our current national security strategy is almost certainly unrealizable in its assertion that "democracy"—the social and political equality of a people under a legal and economic system that enshrines individual liberty and acknowledges their sovereignty in governmental matters—is a natural and wholesome yearning, and that the various forms of tyranny, if not always in opposition to human aspirations, are unable to compete with democratic movements that are fortified materially and morally by the United States.

Here's what's wrong with this outlook as a foundation for diplomatic policy and the military strategy that serves it: human beings are not always governed by reason, even when their own self-interests are at the stake. This is especially true in regard to political matters: people will exchange, with pleasure and in good conscience, what we view as an enlightened idea, for a tribal end. So much has been written about the problems of installing democracy in the Middle East that no discussion of the point is necessary here. But one example of the folly of assuming that democracy can take root once the political machinery is installed can be found in something as apparently irrelevant to strategy as traditions of marriage. As Stanley Kurtz, a fellow at the Hudson Institute, has pointed out, polygamy is widely practiced in the Muslim world, which has an immense impact on the prospects for democracy. Social structures built around polygamy emphasize group solidarity and authoritarianism, at the expense of the dignity and autonomy of the individual. For Americans this is not a theoretical debate. During the nineteenth century, the establishment of polygamy in Utah—a development that the authors of the U.S. Constitution could not have foretold—threatened to spread to other western territories. The issue crystallized before the U.S. Supreme Court. The decision of *Reynolds v. United States* (1878) upholding antipolygamy laws was significant, primarily because it drew a distinction between religious beliefs, which are protected by the First Amendment, and social practices that might spring from those beliefs (such as polygamy or suttee, the Hindu custom of immolating a widow on her husband's funeral pyre), which are not. But just as important, the court's decision "defends the idea that American democracy rests upon specific family structures, which are legitimately protected by law," and that the court's legal reasoning "justifies prohibitions of polygamy by grounding them in a compelling state interest in protecting the social preconditions of democracy." In order for democracy to establish itself in the Muslim world, its citizens must come to terms with issues of this kind and reach an enduring social consensus on them, and that consensus must ultimately take the form of law and established custom.[41]

Examples of the antagonism between a nation's common historical memory and political abstractions are not hard to find among nations culturally close to our own. The rise of socialism during the latter third of the nineteenth century led many political leaders and cultural observers to believe that the laborer's fealty belongs to an idea and not to a specific political authority. Socialism was cast as a globe-girdling movement, complete with its own fraternity (the "International Workingmen's Association"), whose membership comprised representatives from nearly three dozen nations across three continents. Their objective was as ambitious as it was clearly understood: the elimination of capitalism and its acolytes (i.e., property owners) by class warfare. International socialism even had its own slogan: "Workers of the World, Unite!" Nevertheless, World War I, in its uniquely grim way, proved that culture and not abstractions—no matter how apparently beneficent—hold sway. The declaration of war in 1914 ignited among European laborers "no dissent, no strike, no protest, no hesitation to shoulder a rifle against fellow workers of another land," writes Barbara Tuchman. "When the call came, the worker, whom Marx declared to have no Fatherland, identified himself with country, not class. He turned out to be a member of the national family like anyone else. The force of his antagonism which was supposed to topple capitalism found a better target in the foreigner." We seem not to have reckoned with the passions that might blot out the advantages that democracy offers, or which can make democracy into a tool of the very ideas—revanchism, ethnic supremacy, regional hegemony—that ultimately threaten our national security.[42]

The idea that democracy is a fixed and universally understood idea, or that the concept of democracy can be encapsulated in a tightly argued theory, in much the same way as the laws of physical science can, is contrary to fact. The United States is certainly democratic in its legal institutions and culture. But so is Great Britain, which operates under a constitutional monarchy and a legal system that is not written, but based largely on precedent. India is a federal republic with a bicameral legislature, judicial review of legislative acts, and other features drawn from the English constitutional tradition. However, separate personal law codes apply to Christian, Hindu, and Muslim citizens, and fewer than half of the adult female population are literate. The Republic of Georgia is democratic: adult suffrage is universal, the president is elected by popular vote for five-year terms, and legislative and judicial institutions coexist with the executive branch. And the contemporary world contains a number of despotisms that are nominally democratic or republican, because the autocrats in charge, acting in concert with sham parliaments and courts, claim to represent the national will. Browsing the *The World Factbook*, produced by the Central Intelligence Agency, leaves one with the idea that, apart from Iran (theocracy), Saudi Arabia (monarchy), and North Korea (dictatorship), parliamentary democracy rules the world.

The point here is that democracy in the abstract is not what matters—it is a concept that resists the derivation of a coherent theory—but the culture that summons it into being and the mechanisms that make it work. Democracy cannot be mimicked in the same way that a firm buys a license to make products originally designed by a competitor. Democracy cannot be imposed from without; it can't even be created solely because a critical mass of people within a cartographical area desire it—witness the many failures in South and Central America over the past two centuries, as well as in pre–World War I Russia, post–World War I Poland, and the present-day Balkans. For democracy to take root a number of factors must come into play, the most important of which is that there must exist a near universal sense of citizenry, which entails a common language and historical experience, along with a general willingness to compromise and to live peaceably under political decisions reached by consensus. A vibrant democracy depends on a unified culture that can only develop over time, perhaps centuries; it cannot be manufactured or installed.

In his commentary on the French coup d'etat in 1851, Walter Bagehot remarks that "politics are made of time and place," and that the critical element in working out political problems is a distinct national character. Bagehot goes on to argue that 1848, the year of political turmoil across Europe, illustrated this idea:

> In that year the same experiment—the experiment, as its friends say, of Liberal and Constitutional Government . . . was tried in every nation of Europe—with what varying futures and differing results! The effect has been to teach men—not only speculatively to know, but practically to feel, that no absurdity is so great as to imagine the same species of institutions suitable or possible for Scotchmen and Sicilians, for Germans and Frenchmen, for the English and the Neapolitans. With a well-balanced national character (we now know) liberty is a stable thing. . . . [T]he best institutions will not keep right a nation that *will* go wrong. Paper is but paper, and no virtue is to be discovered in it to retain within due boundaries the undisciplined passions of those who have never set themselves seriously to restrain them.[43]

In his appreciative survey of the English constitution (a great book that makes for an engrossing read nearly fifteen decades after it was written), Bagehot attributes the florescence of English political institutions to the national temper: phlegmatic, fair-minded, suspicious of radical change, patriotic. If one wanted truly to understand the anatomy of democracy in a given country, one must first grasp its genealogy.

Wide reading in the history of England (start with David Hume, Henry Hallam, Macaulay, Winston Churchill, Francois Guizot, and C. V. Wedgwood) would illustrate the evolutionary—though not inevitable—progress of democracy among the English-speaking peoples. A dispassionate and intelligent understanding of the American experience would ratify the idea that democracy must spring from native soil. Alexis de Tocqueville's *Democracy in America* makes this point emphatically enough, as does a substantive knowledge of our own founding. Our secession (revolution is an inapt term) from Great Britain was initiated by a desire to recover rights that the colonists, as English citizens, believed were being eroded by parliamentary action.

That ordered liberty and a rationalist understanding of politics are assumed to be unqualified goods is a recent phenomenon. The institutions and societies in which this idea finds authentic expression were half a millennium in the making, and their ability nowadays to sustain themselves in the face of resurgent atavisms and a variety of centrifugal tendencies can appear doubtful to the historically minded. We live in an age in which the wellsprings of national consciousnesses are decaying in Europe, but also, to a milder, but no less worrisome degree, in the United States. We seem paralyzed to thwart decay within our own borders, yet we quixotically and blindly act as if we are capable of inspiring democracy in a region where all social forces—including cultural memory—can be counted upon to oppose it.

Democracy in contemporary America bears only a remote resemblance to the idea embodied in the U.S. Constitution and described in de Toqueville's *Democracy in America* (1835, 1840), James Fenimore Cooper's *The American Democrat* (1838), and James Bryce's *The American Commonwealth* (1910). Since the presidency of Franklin Roosevelt, the relation of the federal government to the citizenry has changed almost beyond recognition, from that of a guarantor of individual rights and the defender of national sovereignty during the first century of the country's existence, to a hyper-busy adjudicator of special pleading by a multitude of interests and, often, an omnipresent and hectoring regulator of private life. One cannot sell or buy a house without completing a form put out by the Housing and Urban Development agency; labels on ordinary household products warn that failing to follow the directions violates federal law.

In the United States, there no longer exists a common understanding of what constitutes citizenship. Or perhaps it is more accurate to say that citizenship—a reciprocal relation between duties and rights—is yielding to residency, which entails free schooling, medical care, and other usufructs of modern compensation culture as the desiderata of immigrants. Attempts to establish English as the official language—absolutely critical given our large population of nonnative speaking, first-generation immigrants—never find expression in legal code. As George Will points out, bilingual ballots are common, thus suggesting that we no longer require that voters "can read the nation's founding documents and laws, and can comprehend the political discourse that precedes the casting of ballots." Immigration laws have proven to be toothless, if the millions of foreigners who continue to work in the United States illegally is any indication. At the time of this writing, massive demonstrations are taking place in New York City and Los Angeles to oppose pending legislation that would make it a felony to be in the United States without proper immigration paperwork. Most of the flags waved by the protestors represent countries in Central and

South America, though Asian and European national flags are visible as well. The one flag conspicuous by its absence in the demonstrations is the Stars and Stripes. This sort of public refutation of loyalty to the United States would have been unthinkable fifty years ago.[44]

Separatist groups have traditionally occupied the fever swamps of the American political landscape. But recently, the founder of the *Domino's Pizza* chain, Tom Monaghan, funded the establishment of a town in Florida that is to be run along strict Roman Catholic principles. One of the motivations was Monaghan's belief that Western societies no longer offer a vigorous religious alternative to radical Islam. The founding of Ave Maria, Florida, cannot be dismissed as a reincarnation of Walden Pond—an attempt to realize primitivism as a rebuke to modern society. On the surface Monaghan's town will be wholly up-to-date—complete with national retail shops, hospitals, cable television, a university, and a governmental administration that shares most things in common with ordinary American towns. Opposition to Monaghan's town springs from environmental monomania (the town is being built on land formerly used for agriculture) and separation-of-church-and-state zealotry. Nothing has been said about the overarching cultural significance of the enterprise. Ave Maria attests to the expiration of a unifying American culture that once sought to transcend race, creed, and social position. The distinctive American identity is thus diminished to a murky shadow of its former self, embodied in governmental activity that centers on the mutation of privilege into civil rights, the issuance of sound money, the collection of taxes, and shielding inhabitants against foreign enemies. Assimilation, to the extent that the term is understood, is no longer a civic aspiration.[45]

The mere mention of a military draft, which unites each generation by providing a common experience that transcends differences imposed by region, economic class, and creed and, of greater importance, emphasizes the noble obligation to serve the state, is opposed not only by the citizenry, but by the Pentagon as well. The populace sees compulsory service not as a praiseworthy thing, but as an obnoxious and extraneous intrusion on one's civil rights. The armed forces are all too happy to concur, but for different reasons. The military departments want the last word on who enters the ranks, which is a wholly understandable point of view, but this view reflects the idea that the services look out for their own interests, even when doing so comes at the expense of a broader national good. Even at the most exalted heights of American politics, one notices the sanctity of national feeling conspicuously eroding. Two former presidents and one former vice president, apparently without compunction and in blatant disregard of precedent, have addressed international audiences about what they perceive to be the folly and iniquity of contemporary American state policy.[46]

Another solvent to our democratic heritage is the absence of a rigorous and coherent understanding of history, our collective memory. One would think that the United States would embrace the idea, given our remarkable past (our achievements in the face of timelessly daunting problems lend themselves easily to fable), but to believe as much is to disregard reality. American history is barely taught in primary and secondary schools, having been replaced by an amorphous miscellany of mostly dubious information called "social studies." Baccalaureate degrees are conferred without any stipulation that graduates demonstrably know anything coherent or substantial about American history. For the most part, historical works found in the major book retailers occupy a range between simpleminded popular fiction, and slapdash narratives assembled by clerks in the service of celebrity scholars.

It is possible that a casual observer of popular culture might think that the United States is awash in a reverence for the past, but this would misconstrue a keen appetite for kitsch and sentimental entertainment for an authentic historical sensibility. We have, for example, a TV channel nominally devoted to history. Hollywood's historical movies (always popular) are for the most part intellectual cotton candy: all sugar and air, in which costume and casting are the main attractions. The material rates as first-class entertainment, but viewers are offered not much more than a collage of pictures given the thinnest of meaning by bits of chatter. Museums are everywhere, displaying everything from valuable artifacts to motley collections of humdrum items from past generations: butter churns, telegraph machines, wood toys. In Williamsburg, where the author once lived, you can buy umbrellas decorated with crudely drawn figures dressed in eighteenth-century garb, and board games that feature colonial themes. One particularly expressive example of our desiccated historical intelligence is the vitality of the nostalgia industry, in particular the prevalence of advertisements that tell prospective buyers of artifacts (or knockoffs of artifacts) that acquiring such things allows them to "own a piece of history." This is a patently foolish notion. History is a uniquely human concept that reflects our ability to organize and interpret past events—a story conveyed by words on a page—and to reflect upon them.

Other bits of evidence suggest the decay, if not quite the abolition, of our faith in self-government. In a majority opinion written by Justice Anthony Kennedy (*Roper v. Simmons*, 1 March 2005), the Supreme Court of the United States validated the use of foreign and international law in constitutional jurisprudence, the implication being that the U.S. Constitution and domestic legal precedent are no longer adequate or even relevant to cases that the court agrees to hear. The opinion was not at all a surprise, as several justices had previously endorsed the idea in public speeches.[47]

One also detects the waning of national consciousness in the increasingly cosmopolitan outlook of the business world, in which the seeking out of markets and attractively priced labor trumps national strategic interests. In early 2006, for example, the Internet portal Google agreed to collaborate with the Chinese government in censoring search results that run afoul of what the communist regime judges to be permissible. The academic profession has drifted in a similar direction. The transmission of America's heritage has been eclipsed by an obsession with multiculturalism, which is commonly a vehicle for hatred of the West. It is thus not hard to conclude that our current NSS sets standards for the rest of the world that we ourselves have difficulty maintaining.[48]

The 2004 version of The National Military Strategy (NMS), which is derived from the National Defense Strategy—a product based on the National Security Strategy—epitomizes our dislocated understanding of strategy. Much like the other strategic documents upon which it is based, the NMS reflects the outlook and relies on the idiom of the marketing professional, whereas consultation with diplomatic and military history seems to have been assigned a low priority, or no priority at all. The concepts and terms that fill our strategic documents are too closely allied with the world of commerce, which bears no vital relationship to waging war. Even more disquieting is that the NMS—as well as the NSS, NDS, and the like—betray an indifference to hard fact in much the same way that a promotional campaign tries to establish a dubious connection between the wares for sale and happiness or wisdom. Note the extent to which advertising agencies strive to transfigure ordinary items (a wristwatch, undergarments, a pickup truck) into embodiments of consumer vanity. Drive this or wear that and youth, romantic ardor, esteem—choose your fantasy—will be yours! Advertising hyperbole does no great harm, and probably works as an indispensable lubricant for commerce. But the frame of mind of the marketing executive can be dangerous when it infects how we prepare to interact with the world militarily.

A useful point of departure is to consider how the military officially defines "strategy." According to the *Department of Defense Dictionary* (2006), strategy is a "prudent idea or set of ideas for employing the instruments of national power in a synchronized and integrated fashion to achieve theater, national, and/or multinational objectives." This definition cannot be reconciled with reason or experience. "Objective," for starters, is a tactical term that connotes a geographical point or a physical thing; it's not a synonym for "a better peace," which is the aim of all sound military strategy. A strategy yields objectives, not the other way round. You don't begin with a set of objectives—seize Normandy, Paris, Antwerp; cross the Rhine so that you can invest the Ruhr Valley before driving on Berlin—and then based on those objectives decide to expel Nazism from Western Europe, while conceding all territories east of the Elbe to Soviet Russia. Rather, you work from the strategic end—unconditional surrender of Nazi Germany—and then decide between alternative sets of objectives; for example, should we attack across a broad front, as previously described, or make a stiletto strike at Berlin, from the lower Rhine? What are the risks and advantages of each option, evaluated by the lights of the strategic end you are seeking to realize, and the means available to you and your enemy?

Of much greater importance is that strategy should amount to more than merely an "idea." Rightly understood, strategy is a closely argued plan to deal with the world as one finds it. The strategist takes a hard look at the facts and evaluates them by the lights of a nation's permanent interests, which for us center on free trade, national sovereignty, and U.S. strategic predominance. Our strategic documents more or less get the latter part right, but otherwise they say nothing substantial about the threats to our strategic position, nor do they offer clear direction on how to parry them. At the end of the day, they amount to an inventory of talking points.

Commercial-mindedness, which comes naturally to Americans, meaning its limitations are largely invisible to us, has cramped our understanding of military strategy. The NMS, the NDS, and the NSS each serve a discrete end, but what they all have in common is the feel of a marketing campaign. They bear little relation to traditional strategic thought. There are notable exceptions to this among our doctrinal publications, of course. "Warfighting" (MCDP-1), published by the Marine Corps, embodies more than 200 years of the best British, German, and American thought on the subject. Each paragraph speaks volumes. But in our major strategic statements we don't even recognize war for what it is: a contest of wills settled by violence. It is the ultimate example of man's struggle with his environment as well as himself: his frailties, irrational ambitions, and the like. Rather we see war as just another entrepreneurial act—a metaphor for a business transaction—and this will do us no end of harm, arguably, it already has.

Browse the thirty pages of the NMS and you will not find a concrete, disinterested assessment of a current or prospective enemy. The document is filled with abstractions about types of threats (the bureaucratic urge to classify is on display), but otherwise the NMS is destitute of any instruction on how to advance or protect our permanent interests, and not a substantive word is said about what sort of peace our enemies and rivals hope to achieve. The NMS is not a strategy at all, but the leading edge of a promotional campaign, the object being to convince ourselves—certainly our enemies cannot have reason to lose much sleep over what's said here—that we are in charge of events, and that we are capable of imposing our will on all menaces. In a word, the NMS presents a disturbingly benign vision of reality.

The cover page of the NMS epitomizes the intellectual flimsiness of the document. Here one finds the slogan, "A Strategy for Today; A Vision for Tomorrow": a rhetorical nullity that makes sense only when viewed by the lights of marketing, in which case the line falls under the promotion rule that you need to come up with a "phrase that pays." The rest of the document can be described by the other four of the five "P's" of marketing. *Product*: suitable deference is paid to technology in the NMS; *position*: for example, how does your business identify itself ("protect, prevail, prevent"); *people*: "joint leader development"; *price*: "force design and size." By itself there is nothing in the NMS that is dishonest or fraudulent—its earnest optimism can be uplifting—but it is wrong to call this document a strategy, if only because it amounts to little more than a collection of hackneyed expressions. There is little here for the military officer who wants to learn about our strategic position, and no factual material to spur thought on how it might be sustained or improved.

A proper national military strategy would evaluate the world unflinchingly by the light of America's permanent interests, and then reconcile strategic ends with available means. On the matter of strategic domination, for example, a reformed NMS would dilate upon the military ways and means that are necessary if we are to transform Iraq into a democratic nation devoted to free markets and individual liberty. America would have to place Iraq under its tutelage for many years, much as we did Germany and Japan after the Second World War, which would require a comparable measure of military and diplomatic power along with the necessary political will. We would have to secure Iraq's borders, take responsibility for its foreign policy and national defense, make sure its financial and civic systems functioned efficiently, and, most importantly, shape the political culture to a degree not seen since General Douglas MacArthur ruled Japan.

The first step would be to grant legitimacy to only two or three political parties, because such a system is the best way to establish political moderation. We would have to make sure that the parties were based on political ideas rather than sectarian or tribal loyalties. Cultures unfamiliar with ordered liberty tend to generate political groupings that dominate and suppress the opposition; elections determine which mob rules the streets and which is persecuted. Collaterally, a prudent military strategy would work from the assumption that Iraq will never stabilize unless the current Iranian government is replaced by leadership that finds no strategic benefit in a chaotic Iraq.

There would also be a discussion of the emerging strategic challenges posed by demographic changes in continental Europe that are, if anything, gathering momentum because there is no concerted resistance to them, other than official acquiescence on the far side of the Atlantic. It appears that the *jihadists* are employing the indirect approach against the United States—or exploiting developments in ways that bring the concept to mind. Operations in Iraq and Afghanistan, the persistent threat of terrorist attacks against the homeland, and diplomatic wrangling with Iran, North Korea, and Russia tax our military forces and absorb our political will. Meanwhile, unassimilated Muslim minorities are gaining influence over Europe's political machinery in ways that might encourage European Union statesmen to frustrate or loudly oppose American strategic initiatives—leaving us diplomatically isolated and beleaguered. Politicians in representative democracies are going to answer to the clamors of their constituencies, which across Europe include a growing number of immigrants who have brought with them contempt for the modern West—a habit of mind abetted by multiculturalism in all of its practical manifestations.[49] As Sam Harris pointed out recently in a *Los Angeles Times* column, "The people who speak most sensibly about the threat that Islam poses to Europe are actually fascists."[50] One would only have to take note of the electoral success at the local level of the British Nationalist Party to understand that Harris isn't exaggerating.

Indeed, we ought to pay keen attention to the growing estrangement of the United States and our oldest and most reliable ally, the United Kingdom. In February 2007 the British government announced plans to withdraw half of its 7,000 troops from the Iraq theater. The spin put on this decision by some commentators was that the diminished British force represents progress in pacifying Iraq. But southern Iraq was never the cauldron that other parts of the country proved to be, and in any case, if the troop redeployment did not signal a wilting of British resolve to support the USA, why were the troops not moved to Baghdad, where they are needed?

There are other disquieting signs. The city of London recently sponsored a debate concerning the nature and consequences of militant Islam. The mayor of London, Ken Livingstone, spouted what have become his customarily bilious, anti-American views and, conversely, gave free play to his benevolent attitude to the radical Islamic subculture of London. In a wiser age, remarks of this kind would have been delivered atop a soapbox in Hyde Park. But at about the same time as Livingstone was blaming the United States for the world's troubles, the British government's foreign office forbade the term, "war on terror," from appearing in official communiqués.[51] The political pressure behind this rather craven move almost certainly did not come from the descendants of Drake and Hawkins in the West Country, or the Barbour jacket-wearing villagers in Northumberland or the East Riding. Livingstone's views, and the reality they represent, should come as no surprise to observers of British culture, particularly readers of Melanie Phillips's *Londonistan* (2006).

Not all Britons loathe their past or have climbed aboard the despise-America bandwagon, but this is a fact that may hasten rather than retard the disaffection between the United States and the United Kingdom. George Walden's *Time to Emigrate?* illuminates one of those little diagnostic truths that adumbrates the larger truth. It is of a piece with the reduction in size of the Royal Navy to the point at which it is no longer a strategic force.[52] Walden's picture of contemporary England bears no recognizable resemblance to Churchill's native land or, for that matter, to the United Kingdom of Margaret Thatcher. The novel takes the form of a letter from a father to his son, expressing a grim sympathy with the latter's tentative decision to rear his children outside of the British Isles. The father, a former member of Parliament, surveys England's contemporary troubles and ponders the impending collapse of civil society, fueled largely by the country's self-injurious immigration policies. Though nominally a work of fiction, Walden bases his work on real incidents and statistical fact.

The stoutest cultural and strategic ally of the United States is expiring, and will likely be replaced by a nation that can offer us little in the way of military help, and that might prove, at times, to be passively hostile to American strategic interests. Certainly, we will no longer be able to assume, as we once did, that the United Kingdom looks upon America as a force for good. Relations between the United States and the United Kingdom have not always been genial—up until the Great War we were as often as not rivals—but such difficulties have largely sprung from economic competition, aggravated by chauvinistic patriotism on both sides of the Atlantic. What is emerging nowadays is something far more worrisome and perhaps irreversible.

A sound military strategy, therefore, would consider the impact of what appears to be a genuine, if oblique, menace to America's permanent interests that is gestating between the Volga River and the Irish Sea. Henry Kissinger briefly discusses this in his book *Diplomacy*. Hegemony over Europe by a single power "remains a good definition of a strategic danger for America," Kissinger argues:

That danger would have to be resisted even were the dominant power apparently benevolent, for if the intentions ever changed, America would find itself with a grossly diminished capacity for effective resistance and a growing inability to shape events.[53]

Kissinger's book appeared in 1994. In the intervening years the situation has grown more, rather than less, ominous. The rise of the European Union, with its supranational legislature that is congenitally antagonistic to America's permanent interests, along with the growing influence of militant Islam across Western Europe, and the incipient belligerence of Russia, should figure largely in our military strategy.

For the first time in its short history as a dominant player in the international arena, the United States must adjust to a world in which balance-of-power calculations—similar to those that shaped European diplomacy during the nineteenth century—hold sway. The United States can be expected to bind with other nations to confront a specific strategic threat, and we will continue to work as a part of coalitions so long as our national interests coincide with those of other countries. But it would be a mistake to assume that every threat to the United State's permanent interests will be viewed by other nations in precisely the same way, to where they will incur comparable risks with us to liquidate it. History shows that nations compete with each other far more often than they cooperate. A military strategy ought never to assume, as ours apparently does, that nations prefer to see themselves as gears in a machine designed to serve the strategic ends of America. An up-to-date NMS would work from the assumption that American expeditionary campaigns are likely to be solitary ventures; and the size and structure of our armed forces should reflect this eventuality.

If we are to protect our permanent interests, each of the military departments must be massively expanded—the sea services in particular, given that we are principally a maritime power—because a military force, no matter how technologically advanced, is deprived of a vital measure of flexibility if it is too small to deal with crises across the globe, some of which might take years to resolve. It is dispiriting to note that the one element of our military establishment that has expanded sharply over the past few years has been the various headquarters staffs, even as our manpower numbers and equipment inventories seem absurdly small when measured against the multiplying threats to our permanent interests. We are in the midst of standing up a second combatant command in the wake of 9/11—the Africa Command—but from where are we to draw the forces to execute the missions that might come its way? A proper military strategy would justify developments of this kind and reconcile ends with means.

One plausible explanation for our inadequate strategic documents is America's limited experience as a great power (as opposed to a hegemonic one, as we were during the Cold War). But history can help us. An illuminating example of solid, strategic thinking can be found in General Friedrich von Bernhardi's *Germany & the Next War*.[54] First published in 1912, Bernhardi's book is very much a creature of its time, insofar as it reflects Germany's desire to dominate Europe. Putting aside the national arrogance and (what is in retrospect) the crackpot anthropology that underpins much of the argument, *Germany & the Next War* offers a clearheaded assessment of Germany's strategic situation, materially, demographically, geographically, and politically. The resources and strategic intentions of Germany's principal antagonist, France, are given due attention, as are those of Great Britain, Russia, the United States, and other strategically relevant nations. To read Bernhardi after surveying the current NMS and its predecessors is to glimpse the conceptual faultiness of our understanding of military strategy.

In writing a national military strategy, we ought to view the world with an informed and mildly pessimistic (a synonym in this context for "prudent") eye. This is primarily the duty of military officers, whose ultimate responsibility is to provide strategic advice to civilian authority. If military officers are not willing or able to do this, who will step in to fill their shoes? Increasingly, it seems that strategic ideas come from civilian sources: think tanks, journals of opinion, and the like. Maybe this is so because the intellectual character of traditional strategic thought is at odds with our contemporary military culture. We idolize firsthand experience without taking into account its limitations.[55] Aggravating matters is our penchant for mimicking the business world, the result being that we emphasize efficiency at the expense of an understanding of the importance of a strategic reserve.

Indeed, the NMS presents readers with a parody of "economy of force." Its discussion of "Force Design and Size" emphasizes "optimizing" forces, and seeking "innovative ways" to realize efficiencies. In the business world one tries to provide just the right amount of product to meet market demand; big inventories are costly. This works well if your aim is keeping the local auto dealer stocked in oil filters and timing belts, but it doesn't work in strategy, and it doesn't work in war. Overwhelming numbers of ships, planes, and troops pin down the enemy by threatening him along several avenues of approach, thus keeping him off balance, and giving commanders the option of picking an opportune moment to deliver a decisive blow. Alternatively, strategic reserves provide commanders and their civilian chiefs with a force that can deter enemies and rivals, who might

Lessons Not Learned: Strategy, War Plans, and the United States Armed Forces 51

try to achieve their own strategic ends while the United States is distracted by combat operations elsewhere.

That our national military strategy betrays a nearly complete absence of sound thought is a dangerous development, which is not without parallels to the circumstances that detonated the Great War. In 1914, military strategy—embodied in operational plans that pivoted on mobilization schedules—was out of sync with political and diplomatic conditions. Today, the situation is reversed. In a word, the NMS (as well as the NDS and the NSS) reflect America's lofty aspirations and our love of business idiom, but they have lost touch with contemporary geopolitical fact and a traditional understanding of strategy.

With its enthusiasm for technological and managerial know-how, the *Quadrennial Defense Review* (QDR) is hardly less dismaying. The QDR appears at first glance to be an in-depth analysis of the DOD's progress in fulfilling the obligations set forth in the NMS. But much like the strategic documents that it supplements, windy rhetoric takes pride of place over clearheaded analysis. For example, in its ninety-two pages of text the QDR makes use of nearly five dozen photographs. The images are astonishingly benign, given the subject matter they are intended to illuminate and the military context that produced them. Soldiers in the photographs are seen operating machinery, conversing with local inhabitants in places overseas, conducting briefings, or handing out food parcels. A visitor from another planet flipping through the QDR might very well be left with the impression that waging war is an admixture of gazing into computer screens, ceaseless grinning while one engages in chitchat, and shaking hands. A clutch of pictures is given over to the discharge of weapons, but the captions tell us that they were taken either during training exercises, or test and evaluation missions; they are the raw material for recruiting posters, and not the sort of image one finds in discussions of how to wage war. Two photographs (about five percent of the total) depict combat scenes: one is of armored vehicles on patrol in Mosul. The other, not surprisingly, shows a robot approaching a mushrooming cloud of dust created by the detonation of an improvised explosive device. The caption informs readers that the "increasing use of robotics has improved U.S. force protection significantly in Operation Iraqi Freedom."[56] This is a particularly apt example of the QDR's dislocated perspective: machines can purge brutality and uncertainty from war; self-protection is the principal mission of expeditionary forces.

The photo and caption of the battle-tested robot bring to mind an anecdote that illustrates the confusion that afflicts a military force, which is captive to the sort of thinking more at home in the business school lecture hall. About two years ago a general officer caused a minor flap by remarking, apropos of the current war, that his soldiers enjoyed killing the enemy. Much of the national press drew a psychic buzz from the vapors that wafted up from their own editorial pieces and commentaries, which were thoroughly marinated in 180-proof sanctimony, and pasteurized so as to be free of the slightest awareness of military history.

Responses to the three-star general's commentary from the Pentagon ranged between demurral and gentle reproof—a circumstance that disheartens even as it fails to surprise. Surgeons don't shy away from celebrating the ejection of a cancer, and the attorney justly takes pride in winning a landmark case; yet our PR-sensitive military culture—a consequence of our embracing of the corporate executive's mental outlook—is embarrassed by ordinary observations on how things are in war. The student of military history would take no notice of the general's remarks, and probably could not be bothered to speculate on the banal question that provided the occasion for them. It scarcely needs mentioning that morale among soldiers is never higher than when victory over the opposing force turns into a rout—the visible signs of which are great numbers of enemy dead and long columns of prisoners moving to the rear, while their surviving compatriots fall back in disarray. By contrast, a beleaguered army becomes obsessed with its survival and relies increasingly on appeals to hope and other emotions (witness the German Sixth Army in Stalingrad, December 1942), something that reflects and aggravates the impotence of commanders tormented by events.

The soothing images of the QDR betray a discomfort with the hard realties of war and give license to wishful thinking. Even more disquieting is that the vocabulary—the carrier of ideas—of the QDR is benign to the point of self-deception. "This QDR defines two fundamental imperatives for the Department of Defense," readers are told on page one:

- Continuing to reorient the Department's capabilities and forces to be more agile in this time of war, to prepare for wider asymmetric challenges, and to hedge against uncertainty over the next 20 years.
- Implementing enterprise-wide changes to ensure that organizational structures, processes, and procedures effectively support its strategic direction.

Substitute "competition" for "war" in the first sentence, and the DOD's "fundamental imperatives" are indistinguishable from those of a firm that retails cosmetics. To the extent that "fundamental imperatives," which we might translate as "aim," of the armed forces need to be stated at all, a more accurate formulation might be this: "The military exists to intimidate our nation's enemies and, failing that, to liquidate them once ordered to do so by the President of the United States." At the moment these words are being typed out on a computer keyboard, news agencies are broadcasting that North Korea has successfully tested a nuclear weapon comparable in power to the bomb dropped on Hiroshima—a strategic menace that cannot be apprehended by the wooly terms and inoffensive photographs that constitute the QDR. Jihadists and lunatic despots, disciples of unreason who are destitute of compassion, are motivated at heart by "the anarchic passion to smash." Preparing the armed forces to confront foes of this kind by couching military reform in the patois of marketing and commerce may ultimately prove to be self-injurious.

Such criticism cannot be dismissed as a pettifogging concern with semantic hygiene. The terms used by the military officer are different from those of the business professional, because they each occupy a world that is essentially alien to the other. They use, or should use, a distinct vocabulary, because their occupations require them to think differently. The business executive strives to expand his company's share in a market that cannot function without agreed-upon customs, universally respected laws, and binding trade agreements. The CEO who violates these things will undermine the reputation of his firm (thus delivering a punishing blow to its stock price), and he likely will face litigation at home and the expulsion of his company's products from overseas markets. Hardly less important is that the business executive must take care not to offend; he must master the art of projecting an agreeable image of himself and the goods his company peddles, because sentiment divorced from humdrum reality plays a dominating role in the success of a business. Consumers choose a certain brand of automobile or necktie, or watch a television program, in part, because they are persuaded that doing so confers status or reflects a discriminating taste.

The patterns of thought that propel commerce are most often the catalyst for military failure. Take, for instance, the decisions of the French High Command in 1914. In preparing for the inevitable struggle with Germany, the French Supreme War Council, led by the minister of war, not only rejected a sound plan to parry an attack through Belgium (which they knew would be the German army's most likely avenue of approach), but also deprived its author of his job and reputation. General Michel, the highest-ranking officer in the army, clearly grasped what needed to be done to thwart a German invasion. His enemy, however, was not bold strategic thinking on the part of the German High Command, but an admixture of hubris and emotion-soaked military dogma that controlled the outlook of his civilian chief and most of his peers.

In the four decades that preceded the events of August 1914, French diplomacy and strategy were driven by the desire to restore national honor, which had been terribly disfigured by the terms imposed on France by imperial Germany. But by 1900 the French High Command's faith in the doctrine of the offensive could not be reconciled with geographical, demographical, or technological fact. It did, however, serve perfectly as a vehicle for French resentment and ambition. Above all else, France was determined to recapture Alsace-Lorraine. A widely distributed illustration from the period shows the extent to which the High Command was beholden to a chimerical view of France's strategic situation. In Georges Scott's drawing, victory was represented not as the destruction of the German army, and the obliteration of that country's goal of European domination, but by a damsel (representing the territories lost in 1871) held in the amorous embrace of a handsome French officer.[57] The offensive spirit that carried the authority of divine commandment proved to be nothing more than a toxic conceit that flattered French vanity and, naturally enough, perverted strategic judgment.

Reality obtruded itself within days after the first shots were fired. As German forces poured across the Belgian frontier, French cavalry units seized the Alsatian town of Mulhouse, and were pushed out immediately afterward. Many citizens who celebrated the return of French troops, faced retribution at the hands of German soldiers; so too did the French army, as German-speaking residents sent word back on the strength and disposition of French forces in the area. Had the French army been less bedazzled by the fanciful image it nursed of itself, and acted instead on General Michel's plan to concentrate forces in northern France, the war would have taken a much less sanguinary turn.[58]

Irrespective of time or place, the commander's stock in trade is danger, chance, fog, and friction: forces that cannot be disciplined by the latest idea out of the Wharton School, or enchanted by Madison Avenue wizardry. War quickly overwhelms the PR-sensitive manager's native reluctance to face difficult alternatives. Defeat cannot be meliorated by a slick ad campaign or an amiable Adobe PDF booklet. Chaos is always the resourceful commander's best friend; and as Basil Liddell Hart amply demonstrates, the indirect approach—so long as it is animated by the resourceful perseverance of commanders and the stoutheartedness of their troops—is the surest path to victory.[59] Put another way, the successful commander doesn't obey "laws" on how to conduct a campaign in a manner similar to the business executive operating in the marketplace; he makes new laws by turning conventional expectations on their head. We admire Erwin Rommel, commander of the Afrika Korps—bold to the point of recklessness, blunt to subordinates and seniors alike, given to keeping his own counsel, scornful of grandiloquent murk—but he would have lasted all of five minutes as the head of IBM.

From the age of von Moltke the elder, up through and including the 1930s, there was a belief, resurgent between brief periods of dormancy and still with us today, that legal processes can render war unnecessary. Witness the Kellogg-Briand Act (1928), in which the signatories—the world's major powers and several lesser states—renounced war as a means of settling international disputes. The QDR has replaced one panacea with another. The dogma inherent in the QDR asserts that technology will largely expunge chaos and cruelty from war, and under the adept hand of military managers can transmogrify armed conflict into the likes of a high-powered business project, bereft of the anachronistic barnacles of fog, friction, chance, and suffering. The traditional begetters of war—fear, honor, and interest, as Thucydides tells us—will be subdued not by Clausewitz's conception of the military genius, but by managerial "best practices" orchestrated by computer networks.

The narrative portion of the QDR, rather incongruously given the message conveyed by the many photographs, leans heavily on observations expressed, in order to convey the idea that our current experiences are an unprecedented tribulation; for example, our era is "characterized by uncertainty and surprise," which imposes on us the need for "continuous change and reassessment" if we are to triumph over "highly adaptive adversaries." One might say the same thing about Europe in the seventeenth and eighteenth centuries, or the first half of the twentieth century. Every age is beset by troubles, disquiet, and fermenting violence. Certainty is not a constituent part of the human experience, nor can it ever be. And it is every bit as likely that bureaucracy and technology will aggravate, as much as they might meliorate, danger and suffering. For all of the breathtaking advances in military art and science that attended the early years of the twentieth century, it was the folly of the German General Staff and the combatants' blinkered faith in artillery as the deciding agent in offensive operations that, among other reasons, delivered up social, political, demographic, and military catastrophe. The executors of Schlieffen's plan altered it in ways that nullified its intended effect. The offensive power of an artillery barrage—a term that originated during World War I—required great accuracy when used to support massed infantry attacks, but it was soon discovered that fragile and unreliable communications made this impossible. Even when communication worked, wear and tear on the gun tubes from incessant firing perverted artillery accuracy even more. The objective of an artillery barrage was to allow advancing infantry to reach their objective and protect them once they arrived, but the new technology all too often achieved a contrary effect. Fratricide was not uncommon; on the other hand a barrage might end too early, allowing a machine gun crew to crawl out of their dugout and pulverize a regiment moving across open country.

The QDR betrays an indifference to history and to recent experience as well. Abstractions are used as a means of evaluating potential or actual enemies. We no longer seek to understand a specific foe. Nowadays our attention is drawn to things.

> In 2001, the Department [of Defense] initiated a shift from threat-based planning toward capabilities-based planning, changing the way war-fighting needs are defined and prioritized. The essence of capabilities-based planning is to identify capabilities that adversaries could employ and capabilities that could be available to the United States, then evaluate their interaction, rather than over-optimize the joint force for a limited set of threat scenarios.[60]

Never has this kind of approach been tried before, because the history of war, which must always be consulted when a nation's survival is at the stake, argues otherwise.

The governing reason for having a strategy in the first place is to purposively shape your forces to meet a variety of threats. Whether or not those threats are properly identified and effective, plans drafted to confront them depends on the mind and character of military commanders and their staffs. Technology is a miraculous and beneficent thing when it relieves man of fatiguing labor, shrinks the realm of suffering, and midwifes the extension of knowledge. It becomes an agent of confusion only when it spurs the relinquishment of judgment, enervates the critical intelligence, and gives license to moral sloth. In war, technological advantages can lose a battle as well as win it. And if the strategy that gives purpose to the engagement has been poorly or hastily drafted, then victory can easily curdle into strategic stalemate, or worse. Countries have been triumphant in war mostly because they out-thought the enemy, understood his moral and material substance as well as their own, and also demonstrated a greater measure of political and military will. By contrast, armies esteemed for their technological and tactical brilliance have failed (the *Wehrmacht* comes immediately to mind), because of their ignorance or contempt of their enemies' strength of will and resourcefulness.

Hardly surprising, given the doctrinaire edge of the QDR, is its continued reliance on assumptions that have proven false in recent years. The feasibility of the QDR is based, in part, on the willingness of allies and partners to supplement and sustain our strategic enterprises. "U.S. operational and force planning will consider a somewhat higher level of contributions from international allies and partners . . . in surge operations ranging from homeland defense to irregular warfare and conventional campaigns." This assumption poses great risks—because nations act on what is perceived to be self-interest. This is a particularly vulnerable seam in our strategy that can be attacked by a prospective enemy, who cannot hope to break through otherwise. Why attack the United States directly when one might hamper our ability to fight by intimidating—that is, making the cost of helping the United States more painful than the gain—our allies?[61]

Elsewhere, the QDR acknowledges the enormity of the challenges that face us, but takes insufficient account of the forces needed to meet them. "Supporting the rule of law and building civil societies where they do not exist today, or where they are in their infancy, is fundamental to winning the long war." The document goes on to stress that concepts and methods are the keys to success, which implicitly discounts the argument that a suitably large force, employed by the lights of operational plans that reflect lessons learned in Iraq, is the most crucial factor in realizing strategic aims.[62]

In sum, at an historical moment when America's influence is on the wane (we will never dominate the world the way we did between the Treaty of Versailles and the end of the Cold War), the vitality of our strategic thinking ought to wax. Our strategic documents, whatever their merits, err on the side of self-flattery and might easily be interpreted by our enemies, rivals, and allies as empty boasting. The frothiness of these documents betrays a cramped understanding of a world that remains brutal, dangerous, and beholden to unreason.

The controlling idea behind our current National Security Strategy—namely, that there is a hard uniformity to human nature—is contrary to experience. The NSS and the derivative strategies that shape military organization and methods are vestigial expressions of the Enlightenment idea (as interpreted by the understanding of the technician and manager) that people naturally crave peace and material prosperity, and that political arrangements can be calibrated in order to liberate these characteristics from the fetid prison of superstition. The problems in Afghanistan and Iraq suggest that we are beholden to the false idea that democracy can be imposed upon cultures that are antagonistic to it. This blinkered view informs our strategic thought and is aggravated by the culture of our modern military, which is tasked to write operational plans that square with our strategy. The emphasis on mastering one's branch specialty, our past military experiences, and (most conspicuously) our technical obsessions are pressures that hardly encourage officers to perceive, let alone understand thoroughly, the underlying unity of war, culture, and politics.

Chapter 2

Transformation Ballyhoo[1]

The United States Armed Forces are in the midst of transforming themselves. But are they really? The ubiquitous employment of the term and the creation of the Office of Force Transformation (OFT) cannot help but provoke a skeptical response. To transform something denotes a major change in form, nature, or function. In the past, armies have indeed been transformed. The Japanese Self-Defense Force (JSDF) bears no substantial resemblance to the Imperial Japanese army of the 1930s. One recognizes the nascent transformation of the English army in the clash between Royalist soldiers under Prince Rupert and General Fairfax's New Model Army. In these cases and others like them, "transformation" was not a product of bureaucratic adjustments or the reshuffling of budgetary priorities. An authentic transformation is imposed on the military from the outside, that is, by great and largely irreversible political and cultural change. The defeat of the cavaliers at the hands of Cromwell signaled the rise of the middle-class army embodied in Kipling's "Tommy Atkins," the long-serving professional (a thing made possible by the ascendancy of parliamentary government), as well as the eclipse of the dilettante soldier of private means and the royal authority he served. The JSDF was the outgrowth of Japan's unqualified submission to the Allies after World War II, an army refashioned in the image of the Western democracies and restrained from territorial aggrandizement by laws that have since been consecrated by time.

Today we face nothing akin to what confronted the Japanese and the English. The U.S. Constitution has not been repudiated. Nor have our political arrangements been extirpated by a conqueror, or our social customs superseded by an alien culture. Such are the engines of military transformation, properly understood. What we're seeing today is an attempt to adapt our forces to current circumstances. Military organizations throughout history have engaged in this sort of thing unceasingly. When armies are not renovated with suitable vigor and intelligence—as was the case with the French armed forces in the 1930s—sooner or later they find themselves trampled underfoot.

Pondering the definition of transformation should not be discounted as an exercise in pedantry. Over the past several years the term has acquired a talismanic significance within the military, displacing "Total Quality Management." A handy way of evaluating our efforts is to survey the pronouncements and documents found on the OFT home page. Technological and managerial themes, couched in opaque language and distended prose, claim pride of place. "At its core, our transformation strategy is a strategy for large-scale innovation," readers of the forty-page pamphlet *Military Transformation: A Strategic Approach*, are told.

> More specifically, transformation strategy is about how a competitive space is selected within which U.S. forces can gain an important advantage. The strategy identifies the attributes within that space that will ultimately lead to an advantage for U.S. forces, not only during combat operations, but also in the conduct of all missions across the full range of operations.

Transformation is described as:

> a process that shapes the changing nature of military competition and cooperation through new combinations of concepts, capabilities, people, and organizations that exploit our nation's advantages and protect against our asymmetric vulnerabilities to sustain our strategic position, which helps underpin peace and stability in the world.[2]

These passages bring to mind an elephant giving birth to a mouse. In *Military Transformation*, thousands of words are deployed in order to communicate a rather pedestrian thought: "we need to update our hardware and administration." But there is something more worrisome than the irony on display here, in which a governmental document attempts to trumpet "innovative thinking" and the shucking of obsolescent ideas, by enveloping its argument in flapdoodle of the sort that during the 1940s called into being the term "gobbledygook." The big problem here is that language is used not to move us closer to truth or to sharpen the understanding, but to mask and so protect bad ideas.

It has traditionally been the American way of war to subdue opponents by employing overwhelming firepower, manufactured and delivered by advanced industrial and technological means. Thus our much-touted radical approach to transformation is actually quite conventional. Hardly less significant is that the OFT Web site embodies the distinctively American habit of transmitting information through the idiom of advertising. We look at war not as experience defines it, nor at our potential enemies as human beings whose thinking is shaped by culture and whose motivation is an admixture of fear, honor, and interest. Rather, we turn inward, yielding wholeheartedly to our culture's obsession with mechanical and administrative wizardry. Watch TV for even a brief period and you will see a commercial featuring business people in conference, the subject of which is either administrative or marketing efficiency achieved through technology. This otherwise beneficent cultural trait—Americans love meetings (we invented video conferencing), and we are always looking to transact business more effectively—has displaced, rather than complemented, our understanding of war.

Colorful graphics abound on the OFT home page, competing with the narrative (which is freighted with meaningless abstractions) for the reader's attention. Conspicuous on the home page is a video accessible at the touch of a button, in which a bespectacled man of middle life, dressed in a dark suit and pale red tie, set off against an indigo-blue background, reads from a document infelicitously called, *Implementation of Network-Centric Operations* (NCO). Hyperbole is not rationed. Viewers learn that network-centric operations are part of an emerging theory of war, and like other epochal discoveries, NCO is embellished by mythogenic characters. There is a "father" of NCO—just like the theory of relativity has Einstein, evolution has Darwin, and powered flight has the Wright Brothers. Apparently, the grandiloquence is meant to suggest that the ongoing controversy surrounding the birth of NCO, a thing revered as much as it is widely known, has lately been settled. History has sorted through the multitude of pretenders and determined NCO's authentic parentage. The self-importance prompts the imagination to conceive an encyclopedia entry a hundred years hence: "Some contend that a retired Swiss electrician who moonlighted as a writer of advertising copy, tinkering in his basement, fathered NCO. But the recent deposit of an unpublished thesis in the Bodleian Library at Oxford, making use of heretofore lost PowerPoint™ archives, establishes beyond doubt that an American can rightly claim the palm of begetting NCO."

Apart from the hoopla, the intellectual underpinnings of transformation are a cause for concern, because they reflect our willful estrangement from a sound understanding of war as it has been experienced across time. One is left with the impression that transformation programs are the handiwork of graduate students enrolled in a middle-tier MBA program, rather than the best thinking of military intellectuals in the service of a great power. Take, for instance, the "capabilities-based" idea, which along with organizational reshuffling is at the heart of transformation. According to *Military Transformation: A Strategic Approach*, the capabilities-based concept:

> recognizes the fact that the United States cannot know with confidence what nation, combination of nations, or non-state actors will pose threats to vital U.S. interests or those of our allies and friends decades from now.... A capabilities-based paradigm—one that focuses more on how an adversary might fight than on whom the adversary might be and where a war might occur—broadens the strategic perspective.[3]

Vacuous observations here and elsewhere in transformation literature are placed in the service of a misbegotten apprehension of what it takes to win wars. Note, for example, the false alternatives put forth as a means of justifying transformation. We cannot know whom we'll fight "decades from now;" therefore we need to build capabilities that address opposing capabilities. In truth, no nation has ever been able to "know with confidence" whom their adversaries would be thirty years down the road. In illustrating this point, one can take delight in imagining the following anachronism: a PowerPoint™ presentation, complete with an Adobe PDF booklet, dated Vienna circa 1895: *The Austro-Hungarian Empire: Vision 1920*. For all the rhetoric meant to suggest purposive audacity, one detects in transformation literature a remarkable passivity, a latter day expression of the Maginot Line. We can't know whom we'll have to fight, so the thinking goes, thus we are right to put faith in lavishly funded equipment and buildings, "paradigms" and "processes."

Our capabilities-based approach is an abdication of judgment. It is the duty of strategists to determine prospective enemies, which is a demanding, but not impossible task. Knowing a culture well is the only reliable way of perceiving gestating sources of hostility. If only Western heads of state had taken to heart *Mein Kampf*, they might have recognized—despite their professional instincts to the contrary and the pacific temper of their citizens—that appeasement was a fool's errand. In his book, Hitler makes plain that he did not see the world as Léon Blum, Neville Chamberlain, and the others did—that is, a view informed by a sharp remorse for the wanton destruction of the previous war, fortified by a vestigial confidence in the diplomatic and intellectual traditions of the nineteenth century, with faith in the improvement of man, and the assumptions that had united diplomats and monarchs across Europe, despite their competing strategic agendas. Had they looked at matters disinterestedly, they might have discerned that Hitler's rise to power incarnated the triumph of ideology—a recently matured force in international affairs—over the practical and compromising spirit of statecraft that the Congress of Vienna bequeathed to Europe. Hitler did not concern himself merely with seeking advantages for Germany at the expense of her neighbors, although it is true that a desire to reverse the decision of the Great War energized popular support for Hitler. Rather, Hitler saw matters through the perspective of a man whose youth was spent in the declining years of the Hapsburg Empire, in which the antagonisms between Teutonic and Slavic cultures (given a false respectability by the racial romanticism that developed toward the end of the nineteenth century) figured largely.

It must have been enormously difficult (but not impossible) for Chamberlain and the other heads of state to have predicted that in the last days of the Second World War, the most tenacious defenders of Berlin would not be conscripts or long-serving troops, but formations named *SS Charlemagne* and *SS Nordland*, as well as other units made up of Flemish, Danish, Norwegian, Latvian, Dutch, French, and even one or two British volunteers. They killed, and were themselves killed, not in defense of a fatherland, but for an idea.

It is not an exaggeration to say that transformation documents are besotted with technology. There is nothing substantial anywhere in the literature to suggest that our national security cannot be realized by mechanical and bureaucratic means, even though mechanical devices and bureaucratic machinery are expressions of (and so are bound by) human aspirations and frailty. At one point, *Military Transformation: A Strategic Approach* does acknowledge that technology is not enough and that we need to change military culture, but readers are led to believe that independent thinking is not much valued. A transformed military culture "must foster leadership, education, processes, organizations, values, and attitudes that encourage and reward meaningful innovation."

> Individually and institutionally, holding on to the past is a result of the natural need to define order in the midst of instability. Individuals and institutions tend to follow what they know and do best because past success becomes the safest predictor of survival in the face of uncertainty.[4]

Note the unreflective dogmatism here that equates human advancement with all things modern. "Education," "values," and "leadership" are expected to naturally collaborate with "processes" and "organizations" in worshiping the new and despising whatever is not new. Note also the implicit coercion. It is the DOD's business to scrutinize the "attitudes" of its employees to make sure that everyone has climbed on the bandwagon. Dissenting opinions—for instance, an argument that opposes "capabilities-based" approaches—are neither expected nor desired.

One sees the technocrat's view in the examples given for such transformation as has already been realized. Under the heading of "Making a few Big Jumps," *Transformation Strategy* argues that the invention of the global positioning system (GPS) "changed the military, the Department, and civil society," and that the army has made "a huge jump by combining new technology with innovative operational concepts. In so doing, it changed the character of land warfare."[5] This is nonsense. Civil society is not much different now than it was in 1991; and if it's changed even a little since then, the GPS played no role. Advances in gadgetry do nothing more than enable us to do a traditional task more efficiently: the railway replaced the movement of goods by wagon train; the telephone superseded the post; e-mail nowadays increasingly is used instead of the telephone; the printing press put scribes out of business, computerized methods of typesetting sent the printing presses to the Smithsonian, and so on. The point is that mechanical things don't make us a better or a worse society, unless we invite them to do so. Yet the authors of our transformation literature would have us believe that a tool can do the work of moral and critical intelligence. A computer or a satellite can do some good provided that the operator is wise or sensible, and his intent is benign. A change in the tools of war is of minor consequence, if a strategy is faulty or malignant, or if commanders are deficient in moral and intellectual ability. The *Wehrmacht* employed tanks, armored semitracked vehicles, and infantry equipped with infrared devices in late 1944. German engineers had developed an assault rifle that could sight and fire around corners. Had these weapons been operated with the Germans' customary tactical excellence, and employed in large numbers and in concert with other technological marvels on the drawing board (the successors to the Me-262, the rocket-propelled Me-163B, shoulder-fired antiaircraft rockets (*fliegerfaust*), wire-guided antitank artillery, air-to-air missiles—the war would not have taken a significantly different turn. Hitler's strategic decisions would have proved to be as disastrous as ever, and the absence of a general staff—effectively liquidated in the summer of 1944—meant that if Hitler in an unguarded moment had decided to seek professional advice, there was none to be had.

It's worth recalling also that if there is one consistent theme across time it is that technical advantages are transitory; during war the employment of a new weapon almost always provokes a countermeasure. A telling fact that illuminates the marginal value of technological development in war—which we have mistakenly exalted—is that Carl von Clausewitz has little to say about the subject, though he had to have been keenly aware of the impact of industrial development on warfare. That *On War* barely mentions technology and instead dwells on the constants of war largely accounts for its enduring relevance. In fact, Clausewitz's *On War* is nothing if not a rebuke to trend spotting in masquerade as intellectual resourcefulness. Clausewitz occupies a diametrically opposed position to the understanding that shapes *Transformation Strategy*. For him, psychological and cultural factors are paramount. That the phrase, "war is an affair of the intellect," came into currency among the Prussians goes a long way in explaining Teutonic military achievements from the mid-nineteenth century until the ascendancy of Hitler as warlord.

One might say that such criticisms are beyond reason, if only because transformation strategies express a general outlook; the details are put in place by individual organizations over time. But this proves to be a feeble rebuttal. The controlling ideas behind military transformation are intellectually arid, which becomes apparent when one evaluates the strongest arguments in their favor. The most effective proponent of military transformation is Douglas A. Macgregor, a retired U.S. Army colonel, who is best known for a book he published about ten years ago, *Breaking the Phalanx: A New Design for Landpower in the 21st Century*, which calls for the wholesale reform of the U.S. military's organization and force structure. In his most recent work, *Transformation under Fire: Revolutionizing How America Fights*, Mr. Macgregor goes into some detail as to how transformation—the capabilities-based approach in particular—is supposed to work in practice. His book deserves scrutiny because it comes from the pen of a well-educated senior officer (PhD, University of Virginia), rather than being a product distilled by bureaucratic processes. There are some astute observations here, for example, Macgregor's trenchant criticism of how narrow orthodoxy cripples fresh thinking in the armed forces. Even so, *Transformation under Fire* is less reflective than advocatory, more of a public relations implement than a dispassionate consideration of the subject. There are no philosophical differences between Macgregor's book and what one finds on the OFT Web site.

The argument of *Transformation Under Fire* is suffused with jargon and embroidered with a multitude of charts and graphs and figures— telltale signs of listless thinking given that the subject is the nature and character of war, the most intense form of human activity. Pseudoscientific terms and phrases, more at home in the self-help industry, are used to describe ordinary (or to mask insubstantial) ideas. A few examples make the point well enough: "Spiral Development opportunities," "Forget the old rules," "Integrated," "Information Age," "roadmap," "Rethinking the objective force with a transformational approach," "asymmetric," "achieving synergy in EBO." Even the title of the book is a bit turgid. A more fitting one would be: "Thoughts on Reorganizing the Army."

G. K. Chesterton claimed that he converted to the Roman Catholic faith because doing so kept him from the degrading slavery of being a creature of his time. The rhetorical bunting of transformation literature is akin to the bondage that Chesterton felt impelled to escape; it is an artifact of our age that will become increasingly inscrutable and irrelevant with the passage of time. Transformation-speak brings to mind Edmund Burke's observation about the dangers of repudiating tradition and the accumulated judgment of previous generations, a frame of mind that reduces men to mere "flies of summer." Books written by Carl von Clausewitz, Basil Liddell Hart, J. F. C. Fuller, and Michael Howard transcend the historical moment that called them into being; they are the fruit of a lively and resourceful intellect operating on experience. Most important, measuring the work of Basil Liddell Hart and the others against the productions of a corporate enterprise such as the OFT, invariably proves that authentic innovation, properly understood, cannot be generated ex nihilo by a bureaucracy—irrespective of the ambitiousness of its mission statement, the credentials and earnestness of its employees, or the opulence of its budget and buildings. Genius cannot be actuated by deadlines or inspired by organizational shell games. Substantial innovation is always the work of a single mind driven by a perceptive dissatisfaction with convention.

The intellectual frailty of our approach to transformation finds expression in the misuse of history. When historical circumstances are discussed in *Transformation Under Fire*, they are approached not from the perspective of the well-trained scholar, but in the manner of the trial lawyer or prosecuting attorney. The historian examines all relevant evidence and interprets it in the service of truth, even as he recognizes that his judgment may differ in emphasis from that of his peers. The historian implicitly acknowledges that no subject can be disposed of with finality. New evidence may come to light; an event from an earlier time may possess a relevance to one age that it does not to another. Modesty, accuracy, eloquence, and disinterestedness thus separate the historical sensibility from tendentiousness or mere glibness. The lawyer handles facts in a wholly different way. Evidence that undermines his case is ignored or downplayed. Procedural maneuvers can be used to pervert or dismiss facts. In order to achieve a favorable verdict, the lawyer can exploit a jury's ignorance and prejudices. Specious reasoning can and sometimes does carry the day.

In *Transformation under Fire*, circumstances of World War II are used in just this way. For example, Macgregor attempts to validate the idea that we need to spend lavishly on up-to-date systems, because aggressive research programs, however wasteful at first glance, are the only way to produce equipment that takes into account recent combat experience:

> In contrast to the British and American approaches, the German and Russian approaches to technological transformation were wildly inefficient. As each armored prototype was rolled out, it was clearly imperfect, but fielding several variants until an effective one emerged worked very well in wartime. How many *Panzerkampfwagen* Mark I and Mark II reached combat? How many Soviet T-26s and T-35s saw action? The answer is, relatively few. Yet the Germans and the Russians fielded all of these tanks in large numbers while the British maintained prototypes, and the Americans bet heavily for the sake of efficiency on one inferior tank—the ubiquitous Sherman. The result was that neither the United States nor Britain developed a tank equal to German or Russian variants until World War II was effectively over.[6]

Macgregor uses these historical examples to draw the following conclusion:

> The army transformational model should involve the following: Look forward to the next technology we can exploit that will help. Field it in limited quantities. Play with it. Test it. Develop new operational and organizational and doctrinal modes for it. Find its weaknesses and strengths. Feed that back into building the next capability and iterate. This means going through an intellectually rigorous process of experimentation in order to reach the goal of sustained military superiority.[7]

There are errors of fact and reasoning in play in the foregoing two passages that cannot help but to corrupt the conclusions drawn from them; indeed, almost everything said is false or an exaggeration. Macgregor's assertions are not so much devious as they are misguided, expressing the attitude that obtains among military futurists—a compound of uncritical faith in technology and a belief that warfare can be domesticated by applying the right theory and equipment. Macgregor and other transformation advocates impose on the past the hopes and assumptions that fuel current debates—even though historical events cannot be treated as if they were akin to a buffet lunch, in which facts are lifted out of their context to suit the tastes of the apologist seeking to make a contemporary political point. Hardly less regrettable is the belief—also false—in the single cause to explain historical events. One belligerent in war stumbled, so current thinking on military transformation goes, and the other thrived because technology was either rightly or incorrectly applied.

Macgregor is right to point out that the U.S. Army was burdened by intellectual sclerosis between the world wars. But at the end of the day the decisions taken were basically sound given the historical situation.

In considering for a moment the perspective of military commanders and political officeholders at the time, there was no point in creating systems of complicated design—especially tanks, an offensive weapon—when there was no compelling demand for them and when, officially at least, we were committed to keeping out of any mess European diplomacy might create. It's also worth recalling that an army made up largely of conscripts, fighting a war of attrition, would have little use for sophisticated machinery; what was needed was reliable weaponry that could be mass-produced in good time. A nation congenitally devoted to individual liberty and free markets cannot convert, as if at the flip of switch, to a culture thoroughly marinated in zealotry for military technology. It is beyond reason to argue that the United States, stretched across a vast continent, protected by two oceans, and with little experience in war as the Europeans had known it, should have spent great sums to develop armaments appropriate to a landlocked power surrounded by enemies.

The M4 Sherman, for example, was hardly an ideal weapon, but it cannot be categorized as wholly inadequate. Nor can it be used as an example from the past that can be handily applied to current circumstances. Our mature automotive industry was easily retooled to manufacture tens of thousands of Sherman tanks. Because we entered the war late, and under immense political and diplomatic pressure to fight a decisive campaign as soon as possible, we had no other choice than to play to our strengths: our vast natural resources and our well developed industrial capacity. In a word, the Sherman reflected the strategic and cultural circumstances that called it into being. All things considered it turned out to be an estimable—and perhaps the most decisive—weapon in the Allied arsenal.[8] Fast, reliable, and easy to produce, operate, and maintain, the M4 represented a good combination of firepower, protection, and mobility. The Sherman was better than anything the British fielded at the time. Though the Sherman was inferior in firepower and armor protection to Germany's Tiger and Panther tanks, it's worth recalling that the latter models were not produced in large numbers, and many of them were liquidated in Russia before the decisive battles were fought in northwest Europe during the summer and fall of 1944. What is more, in some respects the M4 was superior to the legendary Russian T-34, widely thought of as the best armored fighting vehicle of the war. Equipped with a two-way radio set and served by a five-man crew, the Sherman could deliver a greater volume of fire, with better accuracy than its Russian counterpart, which relied on a four-man crew (the commander also served as the gunner), and often went into battle without even a radio receiver. Further, its internal layout was crude and a bit chaotic when compared with that of the Sherman.

The Sherman's armor was relatively thin, but the slope of the front plate, which approached fifty degrees from the vertical, compensated somewhat for this deficiency.

Weighing in at a little over thirty tons, the Sherman ended up being well suited to the battlefields of northwest Europe. By the end of the war, it had been successfully adapted to a number of critical occupations—including use as a flamethrower, mine-clearer, munitions carrier, and artillery platform—and upgraded U.S. Army and Royal Army models began entering service in greater numbers beginning in late 1944. Indeed, the British "Firefly," armed with the 17-Pounder antitank cannon, could take on the Panther, and even the Tiger, on something approaching equal terms. The Sherman also proved to be a nearly ideal weapon for combat against Imperial Japanese soldiers holed up in bunkers during our island-hopping campaigns in the Pacific.

It's also worth mentioning that the U.S. Army considered building a heavy tank during the 1930s, but the plan was abandoned for a number of reasons, not the least of which were logistical concerns. Most port facilities of the day, including lifting equipment, could not handle loads beyond forty tons. The Russian and German heavy tanks, moreover, did not need to be moved across an ocean. And even if the transportation challenges of moving heavy tanks between continents were surmounted, the weight limitations of most bridges in Europe, designed to bear horse-drawn wagons, remained an impediment. Scarcely less critical was the fact that the Sherman could drive long distances on its own tracks, or be transported to the battlefield on four-axle trailers. By contrast, the Tiger and Panther models were not mechanically reliable enough to move over great distances, and the Germans possessed neither tank transporters nor recovery vehicles in suitable numbers to compensate for this shortcoming. Even rail transport was not without difficulty—the running gear of the Tiger tanks was too wide, so the battle tracks, fenders, and outer road wheels had to be removed before entraining. It's also worth noting that variants of the Sherman tank, equipped with a more powerful engine and gun, were in service with the Israeli Defense Force up until at least the late 1970s, and they gave a good account of themselves during frontline combat in the 1973 Arab-Israeli War. This is hardly the record of a patently inadequate weapon.

The Red Army did a reasonably good job of fielding dependable weapon systems, but overall the results were hardly worthy of emulation. The Russians were not innovators but quite the opposite: masters in the art of imitation and titans in the craft of mass production. During World War II their standard pistol was a Browning 1911 knock-off. Their rifle was an updated version of a model introduced at the end of the nineteenth century, and their submachine gun was based on

Finnish and German designs. The Red Army's main light machine gun—the Degtyarev 7.62 mm—was dependable, simple, and effective, but hardly the fruit of bold experimentation. Russian artillery differed from its western European and American counterparts, in that a given shell or cartridge could fit different types of guns of the same caliber, for example, a 7.6 cm round could be fired by a howitzer, field gun, or tank gun, but this had more to do with the Soviet pursuit of standardization at any cost, rather than a desire to make pioneering innovations in gun design.[9]

Russian military vehicles were of equal simplicity. Army lorries were largely of American design and manufacture. Perhaps a third-of-a-million Studebaker, GMC, Ford, Dodge, Chevrolet, and Willys trucks were given to the Russians as part of the Lend-Lease program. The ZIS light and medium trucks, which the Russians relied upon before Lend-Lease, were Ford models built under license. As for the tanks Macgregor refers to, the T-35 was a flop—an illustration of the wooden thinking that afflicted the underlying character of contemporary Soviet culture. With its three cannons mounted in individual turrets, the T-35 was well armed but underpowered, mechanically unreliable, and hobbled by thin armor. It did not influence later designs, though to the Soviet designers' credit the T-35 reflected the sound idea that a tank was of limited value if it did not possess an effective gun. Sixty-one T-35s were manufactured between 1933 and 1939. The T-35's period of gestation was leisurely: two years elapsed between the completion of the original design and the construction of the first prototype (the German Panther took less than half as much time to go from blueprint to initial fielding). The few T-35s that managed to avoid mechanical failure on the way to battling the Finns in 1940, or the Germans in 1941, were abandoned or destroyed. The T-35, however, did make for an impressive sight on parade in Red Square, which probably accounts in some way for its production. In fact, Stalin regularly obtruded his untaught opinions into the design and production of military equipment, so one can safely say that Soviet military equipment reflected the whims of the dictator far more than it exemplified innovation in the abstract.[10]

The T-26 was an example of astute mimicry rather than a willingness to test bold designs. This light tank borrowed the suspension and engine components of the British Vickers Type E, thus refuting the idea that the British failed at mechanical innovation. Twelve thousand T-26s were built between 1931 and 1941—even though its battle record was never impressive. The vehicle received its baptism under fire against Japanese forces in Manchuria in 1934. The resulting experience led to improvements in construction (e.g., welding replaced rivets). In this case and throughout the war, the improvements to fielded systems cannot be categorized as "transformation under fire," but rather as limited and commonsensical adaptations based on immediate necessity. The T-26 performed well against the German Mark I tanks during the Spanish Civil War, but Finnish antitank gunners disposed of them with ease, thus illustrating that new technology carries weaknesses that its advocates often don't see. During the first months of Operation *Barbarossa,* the German invasion of Russia, T-26 tanks were destroyed or captured by the hundreds. Production ceased, not because the Russians recognized the obsolescence of the T-26, but because German forces overran the factories. In a word, the T-26 was a mass-produced light tank of indifferent performance and it did not influence later designs, as Macgregor claims. Far from learning from this experience, the Soviets began building the T-70 tank in early 1942, even though in terms of armor and firepower it was no more capable than the T-26. That it carried only a two-man crew hardly improved its combat capability. Production ceased in late 1943 after some eight thousand units were completed.

The splendid T-34 that Macgregor alludes to began rolling out of Russian factories in 1940. This vehicle was the product not of a culture of aggressive technical innovation, but of the Russians' good sense in copying the chassis from the American T-3, a Walter Christie design rejected by the U.S. Army. The equally marvelous power plant owed much to a contemporary Italian airplane engine. The T-34, no different from the German and American tanks of the same period, sprang from the conditions under which it was to be produced and employed. It was reasonably simple to build, and its running gear took into account the absence of good roads in Russia. The armor plate, sloped at sixty degrees from the vertical, was a genuine innovation that has influenced tank design up to the present day. The ergonomics and optics, however, were inferior to those of the German tanks of the period. The firepower of the 7.6cm main gun was superior to anything the Germans fielded in 1941, about equal to that of the Sherman (fielded in 1942), but inferior to the long-barrel 7.5cm cannon mounted on the German Mark IV beginning in April 1942. Produced in mass quantities—as many as 2,000 a month came from the factories in 1943—the T-34 was operated by unlettered men who were capable of enduring extremes of suffering that would astonish the world. Grouped in brigades, corps, and armies, T-34 units were commanded by officers absolutely indifferent to human losses. It was the semi-primitive character of the Russian soldier and the culture that he came from—something reflected in the T-34 design—and the inability of the Germans to fathom the psychological cast of their opponents, which helped determine the outcome of the war. The point here is that the T-34, like many weapons, was an excellent but imperfect tool, which reflected the priorities, aspirations, and frailties of its creators and operators. No different from any other piece of technology, the T-34 cannot be said to have won the war all by itself.

Macgregor also errs in holding up the German army as an example of purposive innovation. The history of the *Wehrmacht* is an example of an otherwise outstanding military force destroyed, because strategic aims were never reconciled with available means. The Germans had convinced themselves that they could overcome, what was by any sensible person's reckoning, an inadequate industrial capacity by tactical excellence employed against what they believed to be morally and racially inferior enemies.

Cooperation between German military and industrial systems did not encourage or even tolerate the inefficiency that many prototypes would have to be produced before a war-winning weapon was created, as Macgregor would have us believe. The wasteful practices of German armament manufacturers were not the fruit of deliberation, but rather an admixture of bureaucratic self-interest and political wooden-headedness that thwarted the efforts of a hardworking and technologically advanced workforce. The weaknesses in the German weapons industry were many; in fact, they bear faint likenesses to our current approach to military transformation. The Nazi infatuation with weaponry meant that just about any idea militarizing a piece of technology had a claim on resources that grew increasingly scarce as the war went on, a conflict that in the wake of the failure of Operation *Barbarossa* pivoted on attrition rather than mechanical virtuosity. It is an expression of Germany's misplaced priorities that their scientific bureaus were tinkering with a multitude of schemes—theoretical things as well as ambitious modifications to existing weaponry—right up until the Allied armies overran the workshops.

The Germans could not mass-produce tanks because of their complicated design, and also because there existed no expertise in assembly-line techniques; the Germans approached tank production as they did the building of locomotives before the war—relatively few units were assembled with painstaking care. Building a tank also required the cooperation of several manufacturers—the engine, the hull, the gun, the optics, and so on, were built in different locations—which guaranteed that their armored establishments would dwindle as soon as the Germans found themselves fighting long campaigns, rather than the lightning battles of 1939 and 1940.

Never was the German Army's tank arm organized and equipped in a manner consistent with the ideas of its founder, General Heinz Guderian, thus exemplifying the futility of relying on bureaucracies to effect transformation. Time and again Guderian was forced to countenance self-serving and obtuse meddling with his carefully conceived ideas on panzer operations. After the defeat of France, Hitler, who achieved a psychic buzz from gazing at a multitude of military symbols deployed across a map, doubled the number of panzer divisions, not by a massive increase in tank production, but by denuding existing divisions of one of their two tank regiments—increasing administrative and logistical elements within the panzer arm at the expense of its striking power. Aggravating matters was that the artillery branch insisted on control over self-propelled guns and tank destroyers, because their troops could not otherwise compete for the Knight's Cross. The absence of unified direction over the production of armored fighting vehicles became an acute problem during the later stages of the war. Self-propelled weaponry—which could be produced more quickly and at far less expense than tanks—comprised one-third of the panzer force in 1943; by September 1944 Germany was producing more assault guns than tanks.

As for equipping the panzer force, parochial interests within the German military establishment prevailed over military necessity and good judgment. The political influence of Hermann Göring, head of the *Luftwaffe,* ensured that the design and production of aircraft received priority in the allocation of resources. In the scramble for limited capital, the *Kriegsmarine's* U-boat fleet also out-muscled the panzer arm, which was grudgingly and inconsistently supported by the army's leadership in the prewar years, a circumstance which found expression in the inferior numbers and indifferent quality of first generation German tanks.

Contrary to what Macgregor says, the German Mark I and Mark II tanks were not intended to be experimental models—if only because there was little money and scant manufacturing capacity and know-how available for such activity. These tanks were all that German industry was capable of producing at the time. Neither tank influenced later German designs; in fact, the Mark I was based on a British model. Inexplicably, the Germans kept on manufacturing the Mark I and Mark II even after experience in the Spanish Civil War against the better-armed Russian T-26 gave proof of their obsolescence. Just about all of these two early panzers, moreover, saw combat in World War II. How could things have been otherwise? When the Germans invaded Poland in 1939 the panzer establishments largely comprised the Mark I and Mark II tanks, because so few of the more combat-worthy Mark III and Mark IV tanks were available. The same held true for the campaigns in France and the Low Countries—there were about 600 early model Mark III and IV models available—though the panzer divisions were fleshed out with large numbers of Czech-designed and manufactured 35t and 38t models armed with 3.7 cm canons, and smothered in rivets that tended to kill and maim their crews when enemy fire sheered off the bolt heads.

The First, Sixth, and Seventh Panzer Divisions are typical examples of the improvidence of the German political and industrial systems. As the *Wehrmacht* reorganized and expanded after the French campaign in preparation for Operation *Barbarossa*, the Eleventh panzer regiment—like every other panzer division at this time, the Sixth Panzer Division had only one tank detachment—was equipped with 239 tanks, but only twelve of these were Mark IIIs. It wasn't until late 1942, after the division had been moved to France for rest and refit (all of its tanks had been lost in Russia) that the division received Mark III tanks. As late as 1941, several of the First Panzer Division's tank battalions were equipped with Mark IIs. On the eve of the French campaign, General Erwin Rommel's Seventh Panzer Division was equipped with seventy-two Mark IIs, twenty-three Mark Is, and twenty-three Mark IVs, the latter armed with a short-barreled 7.5cm howitzer. The tank establishment of the Seventh Panzer was otherwise filled by captured Czech stock.

Of the 3,300 German tanks assigned to the Nineteen Panzer Divisions that invaded Russia in 1941, nearly as many Mark I (410) and Mark II (746) tanks were in the line as were the Mark III (965) and Mark IV (439) panzers. Several Mark IA tanks—smaller, slower, and less reliable than the grossly inadequate Mark IB—were assigned to Rommel's panzer formations in North Africa in 1941. The Mark I, with its turret removed (these were used in fortifications), was adapted for a multitude of tasks including antitank, radio, demolition, ambulance, munitions transport, and medium artillery platforms, and were in service throughout the war. This was not part of a plan, but an act of desperation, a ramification of Germany's failed industrial and military policies. In January 1944 the last Mark II tank was built—nearly ten years after the first production model entered service, and at about the same time as Germany's sixty-eight ton Tiger II, the "King Tiger," entered production. Early versions of the Mark II tank were used until the end of the war, though largely in the performance of security duties in occupied territories. Much like the Mark I, several hundred Mark II chassis were used as antitank gun and light artillery platforms. The "Marder II" and the "Wespe" served on all fronts until Germany surrendered.

For all of their technical virtuosity, German engineers ignored the value of sloped armor until the T-34 drove the point home at an inconvenient historical moment. Walter Christie's suspension system, of which the Russians made such profitable use, was known to German engineers, but never exploited. Christie was obsessed with speed; his designs featured large road wheels mounted on a suspension with a considerable vertical displacement. Christie was less interested in firepower and armor protection, though he predated the German engineers in recognizing the value of sloped armor plate. The D and E models of the German Mark II tank reflected some of Christie's ideas. These tanks were fast compared with Germany's other models (their road speed was above 55 kph; the other panzer models topped out at about 40 kph), but their combat effectiveness was undermined by extremely thin, unsloped armor and an inadequate main weapon, a 2 cm antiaircraft gun. The two-score manufactured before the outbreak of hostilities were withdrawn from frontline service after the Polish campaign. Even so, production of the Mark II D/E chassis continued until 1942. Two hundred chassis were used as the platform for an open-topped tank destroyer, while the remainder (about ten dozen) served as the basis for a flame throwing tank—an odd decision given the fact that armor protection takes priority over speed in close quarters fighting. Casualties were high among the crews that employed these *Flammpanzers* against the Russians. Had the Germans been able to combine the automotive components of Walter Christie's tank design with their own pioneering work in optics and firepower, the Russian T-34 would not have had any claim to being the best tank of World War II.[11] But what the Germans ended up doing was exactly what Macgregor says we should do today—they experimented with a multitude of prototypes during the war, many of which were produced in small numbers and sent to combat units. The Nazi devotion to technological innovation, which from the moment they seized power began to take on a miscellaneous and whimsical cast, not only failed to produce a war-winning item, but depleted Germany's treasury and further stressed her over-taxed industrial base.

Even when the Germans got technical things right, human agency intervened to gum up the works. The well-designed Mark III tank originally carried a 3.7 cm gun, which was effective enough against potential adversaries up until 1938. Heinz Guderian, whose writings and energetic leadership did so much to shape the panzer force, had argued in the mid-1930s that the Mark III needed a 5.0 cm gun, but the ordnance bureaucracy was already producing the 3.7 cm gun in large numbers, and for economy and convenience ignored Guderian's request. The campaign in France proved Guderian right, so Mark IIIs were produced with an L/42 5.0 cm gun (the "L" here means that the gun tube is 42 times the diameter of the bore, making the L/42 cannon 21 centimeters, or 84 inches, in length). Existing Mark IIIs were retrofitted with the L/42 cannon. In August 1940 Hitler, already contemplating an invasion of Russia, ordered the Mark III tanks to be upgunned with an L/60 5.0cm cannon, but the manufacturers of the weapon did not want to retool because they were already producing large numbers of the L/42 gun tubes. In spite of the consequences that often attended disobeying the *Fuhrer*, the manufacturers managed to evade Hitler's order for more than a year—thus does economic self-interest breed a dangerous contumely. The result of this was that the Mark III tanks equipped with the L/60 cannon did not reach frontline units until early 1942—when that weapon was already obsolete because the Russians were beginning to produce the T-34 in large numbers.

And it wasn't just armored fighting vehicles that were hamstrung by the Germans' indiscriminate affection for technology and their vulnerability to the low ambition that flourishes in a bureaucratic culture. The design of artillery is a good example of an appetite for innovation divorced from the practicalities of waging war. German howitzers incorporated a barrel with an increasing twist, that is, the spiral grooves that stabilize the shell in flight after it leaves the gun tube increase as they approach the muzzle. By contrast, British and American howitzers used a constant twist. German howitzers were neither much better nor much worse than Allied guns—and such differences as existed resided in the carriage, the explosive charge and its propellant, the saddle, and so on. But one thing is for certain: a gun barrel with an increasing twist is much more difficult to mass-produce than one with a constant twist.

In the manufacture of antitank artillery the Germans fielded a handful of types, in small numbers, based on the Gerlich principle—basically a tapered gun tube, cylindrical and rifled at the breech, unrifled in the center, and rifled at the muzzle, features that made them an engineering and manufacturing challenge. These tapered-bore guns came in three calibers, the first number indicating the diameter of the barrel at the breech, the other the diameter of the bore at the muzzle: 2.8/2.0 cm; 4.2/2.8 cm; and the 7.5/5.5 cm. Muzzle velocity was around 4,000 feet per second, about twice that of conventional antitank artillery. Armor-piercing capability was nothing short of amazing at normal combat ranges (though the effectiveness of the shot dropped off at longer distances, unlike conventionally designed guns). The 7.5/5.5 cm shot, for example, could penetrate six inches of armor at 1,000 yards. The problem was that the ammunition for the tapered bore guns was made of tungsten carbide—comparable to a diamond in hardness—and proved to be expensive and scarce. The Germans would have been better off producing standard antitank guns and employing their great metallurgic know-how on reducing the weight of the gun carriages, which proved to be a great hindrance on the Eastern Front, especially where for much of the year the weather was vile and the road network a trial for anything other than fully tracked vehicles.[12]

In another instance, it was Hitler who blundered and the manufacturers who acted prudently. Given the scale and ferocity of combat on the Eastern Front the Germans soon discovered the need for an assault rifle that could function as a submachine gun, squad machine gun, or a rifle at the flick of a switch. One manufacturer, Haenel, produced the MP 43 (later renamed MP 44), a weapon of such excellent design that it has influenced infantry weapons in use today. Yet Hitler forbade its mass production, despite its popularity with troops. Why? He thought it useless and uneconomical; he wanted troops to stick with the standard-issue rifle (originally designed in 1898), for which there was ammunition in plentiful supply. In forbidding the production of the MP 43, Hitler also drew on his experiences in the Great War when the rifle was the weapon of choice for soldiers protected by trenches, and faced by an enemy moving across open ground. But the assault gun was more suitable for war on the Eastern Front, where circumstances might require rapid fire one moment, and a single shot the next. Hitler eventually relented, but his stubbornness did not come without a cost to soldiers at the front.[13]

The administrative and industrial parochialism that hampered the German war effort is not without parallels to our own day. The United States armed forces work under two chains of command. The Title 10 chain, named after the U.S. Code that called it into being, essentially provides for the equipping, staffing, and training of forces organized under military departments ,and directed by a civilian secretary appointed by the president. The army, navy (which administratively comprises the Marine Corps), and air force do not wage war independently of each other. This task is handled by the Combatant Commands, which are organized by global regions: Central Command, Pacific Command, Southern Command, Africa Command, European Command, and Northern Command. The other Combatant Commands (which normally support the regional Combatant Commands) are organized by task: Joint Forces Command, Transportation Command, Strategic Command, and Special Operations Command. When the United States wages war, the individual combatant commanders take direction from the secretary of defense, who is advised by the Joint Chiefs of Staff, which is composed of the military heads of each service and led by a uniformed chairman and vice chairman.

Naturally enough, these two chains compete with each other in ways that not infrequently frustrate, rather than advance, national security—something we ought to consider if we are committed to transforming our armed forces. The four-star generals who reach the summit of their service cannot be expected to always place the good of the Department of Defense over the military department that recruited and promoted them, and that provides them with a generous paycheck, as well as the perquisites that attend the highest rank in the service. And there is also the influence of their predecessors, who have supported their climb up the career ladder. "I gave you an excellent military machine: don't screw it up by giving away the store to help the other services." Pressure of this sort no doubt emerges most often when personnel budgets (particularly, general officer billets) and the purchase of major weapon systems are at the stake.

A good example of this can be found within the United States Air Force, which carries a ration-strength of about 280 general officers. This amounts to a ratio of 1:1,200 between flag officers and all other ranks—which means that relatively few of these fourteen score generals, perhaps 10%, fill command positions. By contrast, the German Sixth Army at Stalingrad along with its supporting units—the largest formation in the *Wehrmacht* at the time and only slightly smaller than the entire U.S. Air Force today—was directed by fewer than thirty generals, including those at army headquarters. This comparison obviously is subject to many qualifications, but we ought not to forget that, unlike the U.S. Air Force today, almost the entirety of the Sixth Army was involved in direct combat at any one moment. Bear in mind also that our strategic and operational communications systems nowadays are light years ahead of what the Germans wielded, thus diluting authority to an even greater degree: generals can communicate among themselves and with their subordinates with much greater ease. The effect of this is that many generals who hold administrative positions find it difficult to refrain from obtruding themselves into operations. Thus it happens that low ambition appears in masquerade as high principle, to the detriment of strategic clarity.

One sees the conflicting loyalties between the perspective of the individual military departments and higher strategic priority, with the USAF's decision to buy the F-22 fighter plane (no doubt comparable examples can be found among the other services).[14] The F-22 is a marvelous creation that reflects America's great technological might in ways that few types of weaponry can. But at a time when we have become an expeditionary force with fewer permanent overseas bases, the F-22 appears to be of uncertain value when measured against its cost. Fighter planes need bases in a theater from which to operate. Unless prior arrangements exist (increasingly rare in places outside of Europe), some diplomatic effort is needed to secure overflight and basing privileges. Rivals and enemies may pressure our allies to refrain from granting these privileges, or to severely restrict them. We can no longer assume that our allies, or prospective allies, will see every threat to us in exactly the same way, or even as a threat at all. In such circumstances, carrier-born fighters and ground attack aircraft are the safer bet. What the air force brings to America's ability to wage war, and which no other service can match, is strategic bombing, airlift, refueling, and communications. More so than any other service, the air force's strategic assets can subdue the twin tyrants of expeditionary warfare: space and time. Yet even as the air force stakes its future on the F-22, it continues to offer combatant commanders increasingly frayed transport and tanker fleets. We have far too few B-2 bombers, barely two dozen. The communications career field in the air force is being liquidated, part of the cost-cutting necessary to fund the F-22. Any number of circumstances may help explain this, not the least of which is that the air force for a couple of generations has been run by fighter pilots, among whom strategic calculation and a conceptual understanding of how war is waged does not appear to figure largely. One way to fix the internecine wrangling among the services would be to resurrect Elihu Root's initiative of building a general staff, suitably fortified by a proper joint warfighting university. But the military departments would doubtless fight such a proposal with all the zeal and sanctimony of the medieval church rooting out heretics.

The exaltation of technology also enfeebles Macgregor's understanding of strategy, as the following passage demonstrates.

> In World War II, the U.S. Army's superior tactical and strategic mobility could not compensate for the vulnerability of thin, ineffective armor on the Sherman tank in fights with the Germans. This, in addition to the fact that American tanks and tank destroyers were clearly outgunned by German panzers had dire consequences for the army's offensive striking power.[15]

Here again one sees on display a crippling indifference to the primacy of moral and intellectual factors in war. One of the things an astute commander does—whether or not he brings to battle superior firepower—is to avoid a direct confrontation.[16] The Allies might well have been able to avoid bloody slogging matches in the Reichwald, the Ardennes, the Hurtgen, and elsewhere, and so triumphed sooner—and perhaps given Western heads of state compelling reasons to keep the Red Army on the east of the Oder River—had commanders made wiser decisions.

For example, the Allies captured the port of Antwerp in early September 1944, yet no thought was given beforehand to the nearly fifty miles of the Scheldt Estuary, which remained in German hands. Failure to secure the coastal approaches meant that inbound Allied ships would be cut to pieces by heavy artillery and floating mines. The Scheldt Estuary fell under the responsibility of the British Twenty-first Army Group. Its commander, General Bernard Montgomery, was content to receive his supplies from the Channel ports. In spite of advice offered to the Allies by the Royal Navy and Belgian Resistance leaders, Eisenhower and Montgomery ignored the threat posed by German forces dug in along the estuary. The Allies were aware of the movement of German reinforcements into the area (thanks to British decoders at Bletchley Park, Oxfordshire), but General Montgomery chose not to act in good time and Eisenhower acquiesced even though he knew that Montgomery's priorities did not always coincide with his own. This alloy of obtuseness, timidity, and vainglory proved to be a strategic blunder, perhaps the worst of the entire European campaign. German forces exploited the delay, moving 65,000 troops, 225 guns, and 750 trucks into the areas northwest of Antwerp.[17]

The port of Antwerp, which fell to the Allied armies so quickly that the Germans hadn't time enough to demolish it, was not in operation until 28 November 1944, some twelve weeks after the Germans were pushed out. The failure to seize the Scheldt Estuary when opportunity beckoned needlessly cost the Allies in blood, treasure, and time. General Montgomery, in particular, well knew that the best way to defeat the Germans was by catching them off guard, before battle groups could be organized out of formations shattered by the Allies' overwhelming superiority in numbers of troops and firepower. The Allies often came off the worse in a tactical clash with the Germans, especially when the Allies were on the offense—as they most often were—if only because combat on the Eastern Front had taught the Germans a great deal about conducting a mobile defense under pressure. One cannot predict what might have happened had the Germans been ejected from the Scheldt Estuary in September 1944, but the failure of Allied commanders on this point almost certainly prolonged the war. Advancing Allied armies had to rely on supplies brought forward by trucks, which was akin to filling a bathtub with a soup ladle. The heavy tanks that Macgregor thinks we ought to have employed would only have aggravated matters. Heavy tanks are greedy consumers of fuel; deploying large numbers of them would have made the fuel shortage more acute and more dangerous to the Allied armies strung out across eastern France.

The point here is that the decisions and motivations of a variety of people—heads of state, military leaders, private individuals—need to be studied and fully appreciated if one hopes to understand why armies succeed or fail. The virtues or imperfections of equipment and military organization cannot illuminate current debates, unless one takes full account of the historical context out of which they are taken. The Germans unwittingly collaborated with their enemies—who were themselves hamstrung by self-imposed fog and friction—in bringing on their own defeat. The Germans' love of technology (almost all of their weaponry was brimming with novelty), their contempt for their enemies, and a septic and foolish understanding of anthropology, encouraged and aggravated a strategy ambitious beyond all reason. In order to appreciate the folly and achievement of the *Wehrmacht*, one would first have to acquire a firm understanding of the previous hundred years of German history and that of their enemies, as well as a working familiarity with the biographies of the combatants' senior commanders and statesmen. This would be a starting point. One would then have to study campaign histories, autobiographies of all kinds, unit diaries, and so on, before one could even hope to draw useful analogies to our own day.

In an effort to encourage readers to climb aboard the transformation bandwagon, Macgregor makes a similar mistake concerning the French army in the years before the Second World War. Here, as elsewhere in *Transformation Under Fire,* history is diminished to the stature of bedtime fable, beholden to a simple-minded moral. Macgregor's point is superficially correct: hidebound thinking is the enemy of judiciousness. What he and like-minded theorists fail to recognize is that their own point of view lacks imaginative boldness and is in fact an embodiment of our straitened understanding of war as essentially a mechanical activity. Macgregor begins the section entitled, "Danger: Old Cultures in New Worlds," by attributing the French defeat in 1940 to the resistance of senior commanders to new ideas. The French Army, Macgregor argues, incorrectly assumed that an offensive posture could only be supported by a large number of long-serving professionals. An army built around conscripts was only suited for defensive battle. Because there was no possibility of the French government expanding the size of the regular army, French senior commanders felt they had no choice but to embrace a defensive strategy once the national legislature voted to continue conscription. In the following passage, Macgregor argues that the "Maginot Line was not the strategic solution forced on the French Army by French politicians"—the implicit lesson here being that military officers who dissent from transformation theory are the spiritual brethren of the French generals whose decisions yielded catastrophe:

> On the contrary, there was substantial political support for an offensive military strategy, but French military culture ultimately trumped any argument for an offensive-force posture set forth by elected and appointed French civilian authority. The Maginot Line concept was derived directly from cultural attitudes inside the senior ranks of the French army.[18]

One is struck by the intellectual poverty in a book that promiscuously misuses historical anecdotes to support its argument. Macgregor's lone source for this assertion is an essay that is based on social-science theory rather than on historical facts.[19] The author of the article relies a great deal on contemporary official documents—apparently without taking into account the doubtful objectivity and accuracy of writings of this kind—and on her own published work as well. But she and Macgregor both fail to weigh the work of esteemed military historians such as Alistair Horne, John Keegan, and Basil Liddell Hart. In the world of the cultural theorist, a single cause, a mass societal compulsion, can explain events. The possibility of one person's courage or folly, or the impact of unintended consequences or motives kept obscure, does not figure conspicuously. In the case of Macgregor's source, "organizational culture" impelled the French into a defensive plan that to us seems foolhardy.

The truth is a meatier brew. The study of the ten-month war between France and Germany—or the study of any campaign, for that matter—cannot be made to yield up a single cause, or be structured so as to deliver a pat lesson that can be applied to contemporary situations. It was not, as Professor Kier and Mr. Macgregor suggest, purely domestic calculations or a cultural struggle between a mindlessly tradition-bound army and an enlightened civilian government that produced a defective strategy. The French High Command believed that they were acting prudently, and one can sympathize with their decisions without endorsing them.

The French embraced a defensive posture in the decade before the outbreak of war, because they felt it played to their strengths. Verdun occupied a prominent place in the professional outlook of French commanders. What worked at Verdun—a stalwart defense that exhausted the enemy—would no doubt work again. Senior French military commanders also knew that the Verdun fortifications had been poorly used. Had the French high command properly exploited the system of fortresses the victory would have been far less costly. Finally, the building of the Maginot Line—a technological marvel in its day—appeared to be an exercise in efficiency born of experience. The open country between the Luxembourg and Swiss borders, with Strasbourg as the center point, offered an invader terrain especially favorable to rapid advancement. Building a fortification across this avenue of approach appeared to be the best way to deter German aggression or break apart her armies should the Germans choose to attack.[20]

Macgregor should have also considered Allied plans written after war was declared in September 1939. At the time defeat was not entirely unavoidable; it was brought on in part by the verve and skill of the Germans at the tactical and operational levels, as much as by weaknesses in Allied strategy—defects that only became apparent after the war was lost. The French and British armies spent an inordinate amount of time during the "Phony War" pondering offensive plans. French strategy ruled out an offensive in September 1939, because in the estimation of French military leadership the army was not yet manned or equipped to do so—a not unreasonable decision given the fact that the German generals saw themselves in a similar situation in regard to the invasion of Poland. The Allied land forces commander, General Maurice Gamelin, determined that it would be

1941 at the earliest before French and British forces would be ready for offensive action. Perhaps by then America would join in—the United States having been expected to play the role of a steroid injected into a weary body, as had been the case during the Great War.

As for the possibility of a German offensive against France before 1941, the Allies accepted the likelihood of such a thing, but they looked forward to it with confidence. They took for granted the moral and material superiority of their own forces over the Polish army—should the Germans dare attack in the same way, French élan would be enough to carry the day. Strategy thus became a straightforward matter of determining the most likely spot for a German invasion. The French discounted the direct route from Germany into the heart of France on account of the Maginot Line. To the north, the heavily forested Ardennes region was hardly more attractive to an invader. That left the Flemish plain, which was all the more enticing to the Germans because the neutrality of that country meant that French and British forces would be out of position for several days after the initial attack. Thus, Allied plans to rebuff a German offensive were built around concentrating the cream of their forces in northern France. No other option appeared sensible.

In attributing French strategy to a single cause—an insufficient eagerness to embrace transforming ideas—Macgregor fails to take into account the immense difficulty of planning for war. Might the Allied armies have repulsed the *Wehrmacht* in May 1940? Yes, provided that the Allies had assembled a strategic mobile reserve rather than relying on a linear defense. Neither Macgregor nor Kier seem to have weighed Clausewitz's argument, validated time and again during the Second World War, that defense is the stronger form of war. A stalemate might have been achieved had the Allies taken a disinterested look at blitzkrieg, rather than brushed off its success as an alloy of German recklessness and Polish ineptitude, and provided that the Germans had gone ahead with their original plan, which called for a move across Belgium and into the teeth of the Allied forces. The reasons why the French placed their faith in the Maginot Line (and their defeat, which was in part a consequence of that faith) were multifarious: history, culture, geography, and emotion, as well as politics, all had a role to play. If one insists on drawing a single lesson from this episode in history, it is that we ought to stand back and make sure that we are not acting on the same sort of blinkered outlook that enfeebled the French and, ultimately, their conquerors. We should spend more time pondering why the French acted as they did and refrain from drawing simple-minded lessons from the outcome of the battle.

In summary, our obsession with technological and bureaucratic machinery and their handmaiden, advertising humbug, tends to divert us from the task of strengthening the U.S. military. Reading the literature of transformation, it becomes clear that we are staking our national security on an untested theory, which is at odds with how wars have actually been fought in the past. It is not quite accurate to say that our approach to transformation unwisely ignores history; rather, it is something much worse. In an attempt to validate transformation theory we have broken history to the saddle of an idea that might be compatible with the study of physics or business management, but is wholly inappropriate to waging war—which is at heart a *social* activity. How could war not be considered as such, given the primacy of fear, honor, unreason, confusion, interest, and the always uncertain relation of motivation, objective, and consequences, which can be counted on to ramify beyond what anyone might predict. Hamlet's remark, "There are more things in heaven and earth, Horatio/Than are dreamt of in your philosophy," is a more apt description of war than anything found in the writings of transformation apologists. Material factors are part of the mix, but they take on significance only as expressions of the moral and intellectual characteristics of statesmen, commanders, and soldiers.

More so than anything else, success in war pivots not on material things but on "the brain of an army." We ought to consider serious reforms that in the past have helped armies confront menaces at least as imposing as the enemies we fight today. A first step to rehabilitating our armed forces would be to rebuild our educational institutions and in so doing demand from graduates that they come away from our war colleges with something more than a line on their assignment history. At the very least, an estimable military graduate program would cultivate an irritable impatience with the flummery that largely constitutes transformation literature.

Chapter 3

The Brain of an Army: Building a First-Rate Joint War University[1]

In 1890 there appeared a highly influential monograph, *The Brain of an Army*, written by Spenser Wilkinson (1853–1937). Wilkinson was a staff writer for the *Manchester Guardian* at the time, but all his life he had indulged a keen interest in military affairs. In his youth he had read about the large continental armies—Prussia's wars against Austria (1866) and France (1870) were headline-grabbing events—that were widely thought to be superior to that of his own country, the United Kingdom. After taking a degree from the University of Oxford in 1874, Wilkinson served briefly as an army officer, an experience that convinced him that Britain's armed forces were in need of major reform.

Wilkinson's dismay at the condition of the British army was motivated in part by the hegemonic aspirations of Imperial Germany. The accession of King Wilhelm II to the German throne (1888) provoked in Great Britain an interest in what we call today, rather pretentiously, military transformation. Wilhelm's foreign policy was at heart anti-British—even though he was Queen Victoria's grandson. He hoped to challenge British supremacy at sea by building a first-rate navy to complement Germany's outstanding army. In response, a Royal Commission was established in 1889 to look into reforming the British army by the lights of her most likely antagonist. In 1890 Wilkinson published *The Brain of an Army*, which focused on the organization and operation of the German General Staff, because he believed "it might be useful to the Royal Commission and to the public to have a true account of that institution, written in plain English, so that anyone could understand it." *The Brain of an Army* ultimately was used to justify the establishment of a British General Staff, and its influence spread to the far side of the Atlantic—Elihu Root became one of Wilkinson's more earnest disciples.[2]

Some of what Wilkinson says is anchored to time and place, but the chapters on military education can be read as a reproach to our current practices. In fact, Wilkinson's commentary on the German War Academy puts in relief the underlying weaknesses not only in our professional military education programs, but the understanding of officership that they have come to reflect. "Leadership" is used nowadays as a buzzword; it is not taken seriously as an idea.

In his discussion of the "regeneration" of the Prussian army after its thrashing at the hands of Napoleon in the first decade of the nineteenth century, Wilkinson points out that the cornerstone of reform was military education—and not as with us nowadays a fixation on weaponry or administration. The War Academy founded by Scharnhorst—Clausewitz was one of the original cadre of instructors—differed from existing training programs, in its recruitment of intellectually gifted officers (taking little notice of a candidate's aristocratic connections), and in its objective, which was to "enlarge and extend their military knowledge and to clear and quicken their military judgment." The entrance examination standard, which Wilkinson translates for us, is one that we would do well to consider imitating:

> The object of the entrance examination is to ascertain whether the candidate possesses the degree of general education and the knowledge requisite for a profitable attendance at the lectures of the Academy. The examination is also to determine whether the candidates have the power of judgment, without which there could be no hope of their further progress.

Wilkinson goes on to describe the character of the exam questions:

> The questions set are to be such as cannot be answered merely from knowledge stored up in the memory, and should test the capacity for clear, collected, and consistent expression. [The paper on conducting operations] must consist of a problem for solution, so as to oblige the candidate to make a decision and give his reasons for it. Each candidate must send in an essay written at home on one of a list of subjects announced some months beforehand.[3]

The German War Academy entrance exam is more akin to the essay contests sponsored by the United States Naval Institute than anything that our war colleges might demand. The beauty of this approach is that it weeds out the careerists—the sort of officer who advances in rank by dint of fierce ambition, reinforced by a judicious deployment of smarminess, and ruthless, self-seeking calculation. A parody of this type could be extracted from a pair of Shakespearean tragedies: an alloy of Polonius's glibness and Iago's egotism. Officers of this kind—these character traits are abundantly in evidence in every other profession, needless to say—are indifferent to ideas and are contemptuous of the prudent risk-taking needed to animate military operations. By contrast, officers endowed with a lively mind and an intelligent appetite for reading military history—which presents a record of how armies adapt to the inevitable imperfections of operational planning—would flourish on the exam as, indeed, they would at operational planning and command.

A couple of objections to the exam's practical relevance to modern times come immediately to mind. Given that the exam would be written away from the college, would not the temptation to cheat be nearly irresistible for the unscrupulous candidate, either by getting help from someone else or by plagiarizing? Second, doesn't this kind of exam penalize officers with distinguished service records who don't happen to be good writers? As to the former concern, plagiarism nowadays is not difficult to detect by electronic means, and resorting to it would be easy enough to discourage by calling attention to its pointlessness. The cheater would soon enough be confounded by the intellectually demanding curriculum; admission to the War Academy merely presents the plagiarist with the certainty of failure and an ignominious dismissal. The latter objection goes to the heart of why we need to recast military education along the lines spelled out in Wilkinson's book. Lucid writing is the reflection of clear thinking. It is not an exotic pastime, such as keeping a reptile farm or painting figures on porcelain teapots. The ability to convey substantive thought in writing is an unmistakable sign of an incisive mind, naturally allergic to the windy rhetoric one finds so often in government documents. The good writer instinctually knows that trafficking in clichés and pseudoscientific jargon is to intellectual health what bathing in someone else's bathwater is to personal hygiene. The canard that the ability to write, and intellection in general, is somehow antithetical to the military leader cannot be validated by history, and is nothing more than an excuse for mental stagnation, a deadly liability for the field commander and operational planner most especially. One learns to write by reading a wide variety of quality books and publications over a lifetime, and by a cultivating self-consciousness when it comes to using words, the carriers of ideas. Effective writing also springs from the ability, indeed the eagerness, to criticize one's own work, which is not an easy (but certainly an attainable) mental habit for anyone who wields authority within a bureaucracy.

There are a number of other compelling features of the German War Academy that Wilkinson touches upon, some of which implicitly inform the discussion below on how we might establish a proper joint war university. The point here is that the culture of the German War Academy, reflected in the admissions exam, was thoroughly imbued with a standard of intellectual excellence that a disinterested observer of our own war colleges—and indeed anyone familiar with our approach to operational planning—would find astonishing. For all the money and time that our war colleges absorb, and their corresponding pretensions to academic maturity (many offer graduate degrees), they are failures when measured against the ideas elucidated in Wilkinson's book. A consideration of the faculty hiring and admissions practices helps make this clear.

At our senior war colleges, faculty comprise a mixture of civilian professors (many with military experience) and active-duty officers the rank of colonel, or perhaps lieutenant colonel. The hiring of faculty at war colleges does not fall into any sort of pattern, though it seems clear enough that intellectual ability, as expressed in one's written work or reputation as a scholar, does not figure largely and is probably irrelevant when weighed against an officer's technical specialty. Generally speaking, faculty have little say in determining the content of the lessons they teach or even in deciding on the manner of instruction. The pattern for new instructors seems to be: "Here are the slides, lesson guidelines, and the readings for the day—make sure you cover the material." In other words, our war colleges are less institutions of higher learning, or even professional schools (as that term is conventionally understood), than vocational/technical training centers that measure success by the uniformity and consistency of the outlook which graduates possess.[4]

It's worth considering for a moment the culture that produces many war college instructors, who in turn reflect and perpetuate the status quo. For starters, innovative thinking does not come naturally to the man or woman of middle life, who has spent many years laboring within the military establishment. Officers advance in rank by knowing the mind of the boss and acting accordingly. "Good judgment" by the lights of any bureaucracy entails meeting the expectations of one's superiors and expressing a suitable deference, not only in regard to ordinary military manners, but also in one's habits of mind. Mental initiative in our armed forces is tacitly discouraged, despite what seems like incessant and noisy references to the primacy of "leadership" and "thinking outside the box." A bold, articulate thinker soon becomes known as an officer of unsafe ideas and, by implication, a disloyal one in temper if not in deed, with predictable consequences for one's career prospects. The rather elderly phrases "team player" and "don't rock the boat" maintain their currency despite the ubiquity of the idea of transformation in all aspects of military operations. Like-minded subordinates are the ones advanced most often by senior officers. So, a colonel or lieutenant colonel assigned to faculty duty at a war college, having spent twenty or more years building a career, is not likely to entertain, or perhaps even understand, ideas that depart sharply from those endorsed by a system that has given him or her a good living, and the psychic compensation that attends advancement in rank.

In *The Re-Making of Armies,* Basil Liddell Hart discusses at length the conventional military understanding of leadership in ways that bear on the constitution of faculty at our war colleges. Though published in 1928, Hart's observations on the subject are as true today as they were a decade before the outbreak of the Second World War. In comparing the great commanders across time with the sort that are advanced by military bureaucracies, Hart puts the lie to the idea that experience and seniority are superior to youth and intellection when it comes to planning for and waging war:

> Miraculously blessed is he whose mental horizon, when he has reached a position of command, remains unclouded by the traditions, social and professional, he has imbibed; whose experience has not been bought at the price of some measure of his imagination, mental independence, and receptiveness to new ideas.[5]

The mental outlook described here cannot help but obtain at our war colleges—which should be the one place from which they should be banished. To the extent that a faculty assignment to a war college differs from others, it is generally assumed that such is the swan song of an officer's career.

What of the student population chosen to attend our war colleges? There are no admissions standards at our war colleges apart from a set of military orders. Personnel officers, taking into account the recommendations of unit commanders, and weighing also the decisions of promotion boards, determine who attends, who doesn't, and who is eligible. The presidents of our war colleges have no say in who is offered a place. And even if they did, it is unlikely that the reasoning behind their decisions would differ from the way we do things today, if only because senior military officers—from whose ranks commanders of war colleges are drawn—by temperament, experience, and training tend to share the outlook of the system that promoted them. Our war colleges (in fact, if not always in what is said in public relations literature) are devoted to the perpetuation of conventional modes of thought. Given that academic ability and solid evidence of mental innovation are not weighed in the building of classes at our war colleges, it is hardly a surprise that no one fails. Our war colleges are designed to admit and refine, that is, "finish," organization men and women, creatures deemed to be as reliable and efficient as any high tech machine.

Promotion to the rank of colonel brings with it the obligation to attend a war college in residence. If we were interested in producing leaders, the opposite circumstance would obtain. Successful completion of an intellectually demanding program would amount to proof, though not necessarily the sole proof, that an officer was ready for higher responsibility, staff work most especially. Currently, attending a war college is nothing more than a way station along a career path, like a tollbooth on a turnpike. Academic excellence is nominally rewarded—"Distinguished Graduate" sparkles on officer efficiency reports—but this does nothing more than feed canine ambition, rather than build on what should be a demonstrable and intelligent curiosity about one's profession. That we fund several war colleges, which in turn offer a multitude of courses, bears no relation to the quality of the programs. What goes on, ultimately, is the accumulation of credentials, rather than education.

We are not much interested in professional ability as that term was understood by the German General Staff. Rather, our war colleges are meant to produce effective technicians and accountants, not intelligently audacious military leaders. Bureaucratic skills are of course important. Much of a commander's time is spent making sure the training pipeline doesn't become clogged, that there are adequate supplies of fuel and munitions, that administrative tasks are discharged in good time, and that higher headquarters' queries are given due consideration. But advanced military professional education should strive after something greater than vocational and technical instruction.

The assumption that our war colleges are nothing more than finishing schools finds expression in the recently floated proposals out of the Department of Defense that the duration of the senior in-residence courses should be cut from one year to six months, or less. But leadership in war most especially in regard to operational planning— demands much more than a short course in administrative matters can deliver, as military history amply demonstrates. A war academy should be devoted to cultivating leadership, which pivots on intellectual resourcefulness. Yet, our war colleges seem breezily indifferent to this idea. There ought to be a lively contention between current ways of transacting business and the culture of a war college, but the objective of military education is to make sure that this does not happen.

The professional military journals published by our war colleges are an apt reflection of the institutions that sponsor them. They are largely devoid of lucid thought. The vast majority of articles in these journals dwell either on technical minutiae ("new ways of employing the *Predator*;" "effects-based intelligence techniques"), or are laden with jargon and footnotes that thinly disguise observations soaked in banality. It is also possible that the referees for the articles, and the editors of the journals, are reluctant to buck the system that justifies their occupation. The one military journal that has devoted itself to challenging conventional thinking, *Proceedings,* is sponsored by a private foundation, the United States Naval Institute. The quality of the thought is unfailingly superior to what one finds in the other military journals. An example of the excellence that a proper military journal is capable of producing can be found in the June 2006 issue of *Proceedings,* which features a pair of essays on whether retired flag officers have the right "to publicly question the President's policies and the performance of his lieutenants." Major General John M. Riggs (U.S. Army, ret.) argues in the affirmative, while Rear Admiral George R. Worthington (U.S. Navy, ret.) takes an opposing view. The July 2006 issue featured an essay by Admiral Stansfield Turner that argued for the diminution, and perhaps the abolition, of the U.S. Navy's fleet of aircraft carriers. On the whole, the contributors to *Proceedings* (and their readers, if the "Comment and Discussion" section is any indication) are interested in ideas, rather than mere technical matters (though these are not ignored). Unlike run-of-the-mill military journals, *Proceedings* is equal, if not superior, to periodicals that serve the legal, medical, and other professions.

It might be argued that our war colleges are fine as they are: why should they not endorse and perpetuate those habits of mind that are necessary to manipulate the bureaucratic and technical machinery of the modern military? Because no army can survive long unless it receives a transfusion of fresh ideas, and this is always the product of a lively mind intelligently expressing dissatisfaction with the way we live now. At some point in a career, an officer who has managed to enrich his inventory of military experience by wide reading in the literature of our profession, and thus has demonstrated the mental initiative that underlies military triumph, must be given a formal opportunity to escape the prejudices of convention without the fear that doing so will spell the end of his desire to put innovative ideas into practice. A keen student of military history knows well that most disasters spring not from technical inadequacies, but from mental and moral sloth that has thoroughly marinated the defeated army's officer corps.

We need to reconstitute our war colleges and, correspondingly, recalibrate promotion and assignment priorities to reward intellectual excellence, if we are to live up to our aspiration of transforming the armed forces. Creating and rearranging existing organizations will not produce worthy results, and broadcasting secondhand platitudes such as "we ought to think outside the box" is equally unhelpful. When used by the military or any other bureaucracy that particular expression is sharply ironical, as it derives from the cult of the software start-up that soon overawes established competitors. The image that springs to mind is of a brace of eager twenty-somethings tinkering in a garage near Silicon Valley. In the military, at least, "thinking outside the box" has come to mean demonstrating heart-and-soul fidelity to whatever idea is passed down from on high. There is a generation of officers who, if blessed with even the faintest pulsations of intellectual vitality, understand "total quality management" as proof that even advanced civilizations are capable of being mesmerized by fatuity, as long as it is packaged in the idiom of the life coach who plies his trade on television and in catchpenny books that clutter shopping mall bookstores.

Captain Alfred Thayer Mahan, toiling away at the Naval War College as he expands his lectures into pioneering works on naval strategy, or Major Heinz Guderian, teaching doctrine at a school for motorized troops that is at odds with the way the cavalry and artillery generals see the world, and writing books that embody his pioneering viewpoint, would find little formal recognition in today's military— hypnotized as we seem to be by transformation hoopla, and eager to misconstrue a narrow ambit of firsthand experience as wisdom. It seems a remote possibility nowadays that bold ideas would be given much credence if they happened to come from the hand of a military scholar or a junior officer, made all the more unlikely if the authors of the ideas lacked the credential of time spent deployed overseas. We need to give serious thought to figuring out the best way of producing a generation of intellectually gifted officers who don't shy from contending with their peers (and, as circumstances demand, with their superiors) in the arena of ideas.

Writing in the *Weekly Standard* on the controversy between defenders of Secretary of Defense Donald Rumsfeld and a clutch of retired general officers, who in the spring of 2006 publicly criticized the secretary's handling of the Second Iraq War, esteemed strategist Mackubin Thomas Owens asserts that the arguments of the latter are "based on two false premises: (1) in general—that the military is always right when it comes to military affairs; and (2) in particular, that the things Rumsfeld got wrong in Iraq, the military got right." Owens goes on to assert that there is no evidence that congeries of senior military leaders at the time came up with anything better, or thought that is was necessary to do so. Owens is correct to point out that Secretary Rumsfeld's reluctance to act on advice from officers who argued for a larger force and more time to prepare was, in part, attributable to what was widely viewed to be the contumely of military leaders during the Clinton Administration—though we should not forget the desire to validate transformation theory played a big role. "Rumsfeld believed that civilian control of the military had eroded during the Clinton administration," Owens observes. "If the Army didn't want to do something—as in the Balkans in the 1990s—it would simply overstate force requirements." The essential truth here is that the armed forces work with an antiquated planning method, something driven, as well as aggravated, by our ineptitude when it comes to preparing officers for operational and strategic planning.[6]

Colin Gray gets to the heart of the matter in his survey of how we approach waging war. The U.S. Armed Forces "have committed three cardinal sins against the eternal lore of war," Gray writes.

> First, they have confused military with strategic success. The former is about defeating the forces of the foe: the latter is about using that defeat to advance the goals of policy. Second, the U.S. Armed Forces have confused combat with war. When the regular style of fighting concludes, the war may, or may not, be over. Also, there is far more to war than combat. . . . Third, as an extension of the second point, the U.S. Armed Forces are especially prone to neglect the timeless maxim that war is about peace, not about itself. This, after all, is the core of Clausewitz's teaching. War is an instrument of policy. It is not a sporting event to be approached and judged with regard strictly to its own standards.

Gray goes on to argue to that we must take a broader view of planning for and waging war. Military planners must "remember their Clausewitz and never forget that warfare is a form of political behavior."

> If the war planners receive intelligent guidance and are themselves strategically educated, they will shape their designs for war with careful regard to the probable impact of their conduct of combat upon the quality of peace that is earned by military victory.[7]

Gray here rightly calls attention to the unity of strategy and operations. The trouble is that, to the extent that Clausewitz is known at all among junior and midgrade officers especially, it is in the form of catch phrases lifted from *On War* in ways that pretty much guarantee a limited or perverted understanding of his thought. There is no sign of a Clausewitzian point of view in how we look at our profession, and it is unlikely that, given the technocratic culture of the modern military (not to mention the broader American culture which it reflects) that an appreciation of Clausewitz is seen as compulsory or even desirable. The way to fix what is an obvious and dangerous deficiency is not by coming up with new slogans and bureaucratic machinery. The most important step is to transform, in the proper meaning of that term, how we educate officers.

The best way of strengthening our ability to produce intellectually resourceful officers who are capable of writing war plans of high quality would be to reform our war college system. For starters, we ought to abolish our midgrade and senior in-residence schools sponsored by the individual services. A course of a few weeks' duration for junior officers that focuses on service-specific doctrine would be of some use, but beyond that officers of average or below-average intellectual and professional ability (by definition, most officers) have no good reason to spend a year or even six months at a service-sponsored war college. Such professional knowledge as is needed to discharge the ordinary obligations of their careers can be gained from duty assignments and training courses. Perhaps a basic distance-learning program can be set up. But it is pointless, or at least grossly inefficient, to fund graduate-level professional schools that, at best, cannot offer more than an indifferent education, and that do not illuminate in any substantial way how wars have been waged across time. The individual services exist to train, equip, and staff the armed forces. This is an essentially administrative function. Perhaps one can assert that military in-residence programs are needed to support this kind of work, but it would be a difficult case to argue successfully when put under the harsh light of a cost-benefit analysis. Eliminating these schools would make available the money to create, and the time for suitably qualified officers to attend, an authentically elite joint war university.

What exists today under that name is, sadly, an imposture. The Joint Forces Staff College (JFSC) in Norfolk, Virginia, has drifted away from the noble and historically sound vision that called it into being in 1946. Over the intervening decades it has degenerated into little more than a convenience for personnel officers casting about for a place to put jetsam—a sinecure for military retirees of doubtful academic ability, and for the students a mildly diverting refuge from the demands of the operational world. It is very much a joyless task to call attention to what in truth are embarrassing deficiencies in how we educate midcareer officers bound for joint assignments. But we are all joint warfighters now, engaged in a struggle against a savage, resourceful, and tenacious enemy. It would be a mistake, moreover, to ignore the possibility that in the future we will face opponents who are much better equipped and organized than our current foe, but every bit as remorseless, brutal, and fanatically self-righteous. It is far better for us as a profession to criticize ourselves in the national interest, rather than to be taught no end of a lesson by a latter-day Stalin or Hitler. The willingness of military professionals to steadfastly and unsentimentally pursue excellence is all the more important when it comes to operational planning. History is loaded with examples of armies made up of valiant soldiers and resourceful commanders, who perished on account of intellectually moribund staff work, and decisions made by intellectually timid and dull-witted headquarters personnel. It is thus to our advantage to attack gestating problems, rather than to write a history in which those problems were allowed to reach full maturity.

The reputation of a professional school—whether of law, medicine, business, or military arts and science—rests on two pillars: the accomplishments of its faculty, and the extent to which alumni attribute achievements to what they learned from their course of study. It scarcely needs mentioning that alumni are more likely to do so based on the ability of their teachers to inspire and mentor, as well as to educate them. By this standard the Joint Forces Staff College is wholly inadequate.

As a general rule—there is of course the odd exception—the faculty at the Joint Forces Staff College fall into two categories. In the first are promotable lieutenant colonels in need of a joint assignment, or who otherwise are deposited at JFSC until their 0-6 selection board convenes. The second category comprises 0-5s and 0-6s, apparently sent to JFSC to run out the clock until they are retired from active duty. What plainly is not given weight in faculty hiring is academic temperament, this is, a respect for ideas, an ability to teach, and an intelligent appetite for studying one's profession. There is a fair number of civilian faculty and staff, but this is hardly encouraging given that many of them simply kept their office keys and exchanged their military for civilian clothes. In fact, several civilian instructors formerly served in the grade of 0-6 as deans of the various schools, a blatant example of the absence of probity and judiciousness in the college's hiring practices. The college's senior "professors" are not expected or even encouraged to publish in professional journals or to write books. As we shall see later in this chapter, doing so might well prove to be professionally risky.

The teaching burden at JFSC rarely exceeds delivering off-the-shelf PowerPoint™ presentations a few times a week to twenty or so students, and attending mandatory "faculty development" sessions that would fit neatly into a K-12 teacher training program, but are completely out of place at an institution that portrays itself as "our nation's premier joint professional military education institution." This is not an insult, but a description. The present writer recalls one faculty training session in which attendees were advised on how to efficiently erase a chalkboard in order to avoid creating the distraction of a wiggling posterior. Another lesson was given over to building paper origami—to what end the present writer cannot recall. In a word, ticket-punchers, sun-setters, and time-servers largely make up the faculty of JFSC—a circumstance that can only have a baneful effect on the quality of the curriculum and the intellectual culture of the institution.

Sitting atop the ill-qualified faculty at the Joint Forces Staff College is a ponderous administrative structure. The commandant of the college usually holds the rank of major general. In addition to a chief of staff and a "Director of College Relations," there are at least six deans holding the grade of O-6 or the civilian equivalent, even though only one of the college's four schools—the Joint Advanced Warfighting School (JAWS)—confers an academic degree, and that degree is awarded once a year to about thirty students. This makes the JFSC unique: nowhere else will one find a 1:4 ratio of senior administrators to students enrolled in a degree program. In addition to the dean of JAWS, the Joint Forces Staff College has a "Dean of Information Technology," a "dean" of the college's online education program, a "dean" of "Academic Affairs"—even though the college quite consciously embodies a vocational rather than an academic spirit—and one dean each for the Joint and Combined Warfighting School (JCWS-I) and the information operations school (JC2IOS), both of which offer short courses that center on administrative subjects.

Given the composition of the faculty at the Joint Forces Staff College, it is hardly surprising that the curriculum amounts to an admixture of simpleminded computer simulations, discussion sessions that cannot help but be low-wattage affairs given the inferior quality of the faculty, and pointless day-long excursions that bring to mind elementary-school field trips. The mandatory visits to Yorktown and the MacArthur Memorial, for example, will find the staff college's students—men and women of middle life who are accustomed to spending the day discharging grave responsibility—rubbing elbows with tour bus sightseers and adolescents under the care of their chaperones. Occasional speeches and seminar talks by retired flag officers and diplomats momentarily hold at bay the intellectual sclerosis of the college, but the prevailing dull-wittedness, and what can only be described as a defiant indifference to vigorous thought, restore the status quo ante soon enough.

In 2004 the Staff College was directed to establish a graduate degree program, a decision that reflected the high seriousness with which senior DOD leaders viewed operational planning. The Joint Advanced Warfighting School (JAWS), which awards a Masters Degree in Joint Campaign Planning and Strategy upon the completion of the ten-month course, has proven to be a great disappointment. In fact the school reflects the traditionally drab intellectual spirit of the Joint Forces Staff College: a new program is launched with a great deal of fanfare, but nothing really changes. The mistakes would be easy to forgive if the school's charter was to start from scratch, but the curriculum, and for all intents and purposes the broader culture of the school as well, was based on current practices at the senior service schools.

The JAWS curriculum is a jumble of poorly conceived and sloppily expressed ideas that might very well provoke a capable and diligent student to wonder why the college should have any legitimate claim on his time. Take, for instance, lesson TH6116A, which is part of the "Foundations in the Theory & History of War" block—one of four that constitutes the JAWS curriculum. This lesson typifies the flaccid intellectual character of the JAWS program; it is neither much worse nor much better than any other in the curriculum.[8] Much of the lesson is so poorly written that at times it flatly refuses to make sense, as in the title: "National Security Strategy and National Security Surprise War From Tectonic Shift—A Case Study." Note also the time allotted: in two hours students are expected to "analyze" the evolution of the national security strategies and the campaign plans of the United States and Japan, from before hostilities began in 1941, up to and including the battle of Midway. As if that is not quite enough to do, in the time it takes to watch a feature-length movie or play nine holes of golf, students are also expected to "Evaluate for [sic] other, past, categorizable paths to war"—whatever that means. On what foundation are students to erect their analysis? They are directed to read a handful of excerpts, totaling about seventy-five pages, taken from two anthologies and one historical survey published between 1960 and 1986. It taxes credulity—but it's true—that the lesson developer is no adjunct instructor or emergency hire, but one of the staff college's senior civilian professors.

The JAWS curriculum shares in common with the JCWS-I and -S programs a misbegotten enthusiasm for field trips. These excursions are a poor use of students' time; they are nothing more than a form of tourism. Students spend hours attending briefings that might easily be reviewed online or, better yet, be integrated into the curriculum. Asserting that such trips allow students to see firsthand how staffs actually work is unpersuasive, if only because the students are not members of the staff, but guests for an hour or an afternoon. Attending dinner at Buckingham Palace does not entitle one to claim an essential understanding of, let alone membership in, the royal family. Especially puzzling is that an entire day is given over to a visit to Yorktown and a week to Gettysburg. Yorktown was a minor affair, a petty example of siege warfare. Gettysburg was a battle of encounter between ground forces that bears no compelling or extensive relevance to modern joint warfighting. These engagements are definitely worth studying. But in a curriculum taught under severe time constraints, allocating six days to visiting these places is hard to justify. It would be far better for students to spend that time reading, in their entirety, a range of perspectives on a modern campaign—for example, the opening engagements of World War I, the battle for Port Arthur or El Alamein—with the objective of deepening their understanding of current operations, and broadening their experience by studying the ways in which staff work bears on battlefield operations. At present, this commonsense approach is not even attempted anywhere in the curriculum.

It is difficult to understand how students might be inspired or even instructed—though they might easily be confused and frustrated—by the JAWS experience. "Academic rigor" is supposedly the foundational principle of JAWS, but what is actually imposed on the students is a great deal of busy work, sham work, and trivial work. For starters, JAWS students are not issued books. Nor are they encouraged to buy them; there is no bookstore at the Joint Forces Staff College. Rather, texts are loaned to them by the college library—similar to the modus operandi of your neighborhood high school. Needless to say, it would be a minor financial hazard for students to use their borrowed books in ways that conscientious graduate students normally do: underlining passages, scribbling observations and questions in the margins, carrying the books about just in case circumstances allow for the reading of a few pages, and so on. More seriously, the reading load—while certainly onerous—lacks coherence. Though the course's duration exceeds ten months, the students never read a major work—such as Clausewitz's *On War*—from cover to cover, which is necessary, but not quite sufficient, if one is to grasp, as the JAWS Web site declares, an "author's premises and assertions." Writing assignments aggravate matters. In the "Theory" block, for example, the students glimpse bits of a few influential works, but mostly they read interpretative surveys or parts of surveys; out of this flimsy hodgepodge they are expected to write a "philosophy of war" paper: a task that is certainly toilsome, but not at all fruitful.

Hanging over the heads of the students—not unlike the sword of Damocles—is the requirement to write a sixty-page thesis, even though the faculty that put together the JAWS curriculum are plainly ill-suited for directing hard research, or even for giving advice on how to write clearly and argue persuasively. Hardly surprising is that the examination and evaluation system is no more than a perfunctory exercise. No one fails. Perhaps it would take a stubborn and truculent ineptitude to justify expulsion, but because no one has ever been dismissed for academic deficiency, or is likely ever to be, this is mere conjecture. And why should this be the de facto policy? An otherwise esteemed officer who failed to make the grade (if only because he might advance ideas that challenge convention) would probably bring an unwelcome measure of scrutiny to the staff college's academic practices.

One can sample the intellectual tenor of JAWS by perusing the school's online journal, *Campaign Planning*, which is a menagerie of pretension, triteness, self-congratulatory buncombe, smarmy deference to convention, and standard-issue bafflegab that highlights, rather than cloaks, the absence of clear-headedness. Thoughtful contentiousness, which should be the defining feature of an academic journal produced by a war college, would be strikingly out of place here. If one exercises the patience to read through *Campaign Planning*, one will not stumble across a single philosophical difference between the contributors and conventional thinking. The issues comprise some seventy pages of PDF text, yet there is nothing stimulating or even useful to be found amongst the many thousands of words deployed here; not a single memorable phrase nor a solitary, piquant observation manages to stand in relief against the sea of murk.

Even the library (generously funded, if staff salaries are a reliable indication) is no refuge from the amateur reflexes of the Joint Forces Staff College. The energy of the library staff seems directed at setting up displays, giving tours, printing leaflets, distributing posters, and keeping the book stacks tidy. Any other duties are apparently viewed as extraneous. Witness the library's hours of operation; when classes are in session the library opens at 0700, scarcely an hour before seminars meet. The library closes at 1800, when many students are still participating in obligatory softball or volleyball contests, and remains shut on weekends and all holidays. No reputable graduate school library would tolerate for a second the idea of operating in this way.

Reflecting and aggravating the deficiencies in the curriculum and administration is the septic culture of the Joint Forces Staff College, where a suffocating conformity obtains. Intelligent initiative, on the rare occasions when it flickers up, is snuffed out, apparently because intellectual vibrancy and moral courage in the form of a willingness to challenge received opinion is viewed as a threat to what is collectively understood as a comfortable and safe manner of proceeding. Though National Defense University regulations protect academic freedom, and honorably encourage professional research and publication, at the staff college such practices are in fact looked upon with great suspicion. A personal example comes immediately to mind.

During his time at JFSC, the present writer published an essay in *USNI: Proceedings* that asserted the under-appreciated value of war colleges in effecting transformation—a point of view that can hardly be considered as exotic or even controversial, especially at a staff college. The commanding officer of the college (no doubt encouraged by his staff) issued a written rebuke to the writer for "attacking the military," which in the commanding officer's view was aggravated by the author having attended the University of Oxford under USAF sponsorship, an achievement that the commanding officer in a private conversation dismissed as "time on the dole." The commanding officer spiced his remarks with a dollop of intimidation, asking whether the author had received orders yet for a pending assignment.

One would have thought that such wooden-headedness had long ago disappeared from the mental outlook of senior commanders, if only because of the blood price extracted from our troops during the Second World War on account of such habits of mind. In 1920 Captain Dwight Eisenhower, West Point class of 1915, published an article in *Infantry Journal* that argued for a more resourceful means of employing tanks—something that was all but unknown to the embryonic *Wehrmacht* at the time, though a few junior officers and their mentors were just beginning to give the concept of *blitzkrieg* serious thought. Eisenhower's clairvoyant essay was rewarded by the threat of a court-martial from the chief of staff of the Army. Because Ike was at the time assigned to a line unit, he at least had reason to weigh the possibility of some form of censure; deviation in such circumstances can be costly to one's career. But those who serve on the faculty of war colleges *should be expected* to evaluate with a disinterested and critical eye our current practices, and recommend ways of transacting business more effectively. Who else has the experience, the education, and (supposedly) the institutional support to do so?[9]

For a fuller perspective on the quality of the JAWS program, it's worth taking a look at how things are done in the civilian world. Comparing JAWS to middle-tier professional schools (let alone to the best, such as those offered by the Ivy League and the more esteemed state universities) would dramatically underscore its weaknesses. But even when measured against less competitive programs JAWS falls below par. Take, for instance, the Management Division of Delta State University (Cleveland, MS), where faculty credentials reflect the expectation that instructors are intellectually engaged with their profession through publication and regular participation in conferences. The college has a bookstore, where students buy recently published titles that bear on the various courses of study, thus suggesting that faculty keep abreast of the latest scholarship (and not merely the latest news) in their disciplines.[10] In fact, it is a taxing thing to try to find a graduate program anywhere that, in regard to curriculum and faculty hiring and promotion, operates along the same lines as JAWS. JAWS most closely resembles the University of Phoenix, an accredited graduate institution that offers online degree-granting programs. Apologists for the status quo might claim that such comparisons are unfair, because our profession differs sharply from others. But must "different" mean "worse"?

Whatever might be said in favor of JAWS, it cannot be argued that the culture of the school is compatible with what one would expect of the premier joint staff college of a great power. Certainly it bears no relation even to middle-tier schools that serve other professions.

The Joint Advanced Warfighting School, which should be the crown jewel of military education, if only because it should have been built on nearly a century's worth of experience, is no more than a gargantuan version of the insipid programs offered by the Joint Forces Staff College. Though by itself an embarrassing problem, the JAWS and the Joint Forces Staff College reflect our approach to command and staff work. We seem committed to producing bureaucrats, people who can be relied upon to move paper and follow steps, rather than to think about the worthiness of what's written on the paper, or whether the steps are likely to beget confusion or folly.

Currently, our joint education programs meet the expectations set for them. In order to reform our outlook on operational planning, we need to rehabilitate our professional educational institutions. War—as distinct from combat—remains an intellectual activity. Officers need to be prepared for planning with the same measure of rigor that is to be found in our training programs and the application of scientific advances to weapon systems. The foundational step toward this end would be to build a first-rate joint war university. It would take a couple of generations of graduates to make their way to the summit of their profession before one might expect to see improvements in the operational plans we write, but the wait would well be worth it.

Building an elite joint war university to replace the counterfeit item we rely on today, and ultimately retooling our approach to the constitution of joint operational staffs and the writing of plans that embody strategic wisdom, is not strictly a matter of bureaucratic reshuffling or upgrading equipment. All existing assumptions about professional military education would have to be given fresh and unflinching scrutiny. To begin with, the college should not be located near Washington, D.C., where there is a temptation to bring in guest speakers, take students on excursions that masquerade as "field research," and so on. There is nothing objectionable to inviting guest lecturers once in a while, but having more than three or four over a year's time undermines the coherence of the curriculum and deprives the students of time needed to accomplish solid academic work. As noted earlier, field trips serve no profitable end. Locating the college away from the nation's capital would also reduce the possibility of outside authorities meddling in curricular matters, influencing faculty hiring and promotion, and exerting pressure in ways that serve an end other than intellectual excellence. The ideal location would be near a combatant command, but close to a metropolitan area with a reasonable cost of living, adequate transportation facilities, and some cultural amenities. The college should not be a tenant unit (an administrative distraction), but should occupy its own real estate. The chain of command should flow directly from the chairman of the Joint Chiefs of Staff to the president of the college. Students and faculty should reside on campus, which would offer facilities that one would expect to find at a first-rate graduate school, including a bookstore.

The curriculum, which would take from eighteen to twenty-one months to complete, should correspond to the exacting responsibility that staff college graduates can be expected to bear. One can imagine the objections: operational tempo and manpower shortages argue for reducing, rather than extending, the duration of in-residence programs. But acting on this point of view would be penny-wise and pound-foolish. A proposal to halve the duration of pilot training or ranger school would not be countenanced for a moment, and rightly so. Why, then, do we believe that pursuing excellence in preparing officers to build operational plans—which more than any other factor determines the fate of armies in war—is somehow of lesser value than technical proficiency?

"War theory" should comprise the first block of instruction at our notional staff college, and take about nine months to complete. Major theories of war will be studied, not surveyed (*On War*, for example, must be read in its entirety), and serious thought would be given to the context that produced them. Campaigns that illustrate, modify, or refute a given theory of war would be studied intensely. Memoirs, biographies, histories, and other commentaries should constitute the reading list.

Military strategy would comprise the second block and take about three months to complete. Here, students will read widely in the masters of strategic thought and thus acquire a sound understanding of the relation of war and politics. The final block of six to nine months' duration, would focus on campaign planning: writing a draft plan; war-gaming; submission of a final plan. The war plan should not, as is the case at the Joint Forces Staff College, be a mere test of how well the students grasp administrative matters. Rather, students would be expected to come up with a plan that can serve as the basis for an actual contingency.

War-gaming should be a demanding activity. Students should know about logistical and technical issues, demonstrate familiarity with the culture of the postulated enemy, and evince a solid knowledge of the enemy's political and military leadership. Fog and friction should come into play. Students would be expected to explain the reasoning behind their decisions, and accompanying the final plan would be a closely argued dissenting point of view. This requirement would demonstrate that the students carefully weighed plausible alternatives and reasoned their way to a final decision. Senior leaders of the Joint Staff would be encouraged to review the students' work by the lights of existing war plans. In other words, the plan is not simply an academic exercise, but potentially a contribution to national security.

A word must be said about the elements missing from this curriculum, but which figure prominently in current joint education programs. No time is allocated for "service perspectives." Completing a thesis is not required. Successful applicants to the college will have demonstrated a solid understanding of each service's doctrine and culture as part of the entrance exam, in which answers take the form of essays. The day-to-day life of the college will build on this foundation. Writing a thesis (a research project of considerable length that argues an original point of view) takes too much time to fit within a twenty-month course. The submission of a final war plan squares perfectly with the mission of the college and will require great mental exertion. It is thus comparable to the thesis requirement of civilian programs.

The manner of instruction at the college would also differ markedly from current practices. The seminar method would still be used, but far less time would be spent in the classroom: no more than three hours per day in a four-day week. The students would be issued books, but these would only form a skeleton reading list. Students would otherwise be expected to read widely and bring this enriched perspective to class discussions. Writing assignments would depend heavily on interaction between the individual student and a faculty member. Letter grades (a method of evaluation that is most effective at the high school level and below) would be replaced by narrative evaluations. Advocates of our current manner of proceeding might argue that such a system lacks the rigor that only coercion can deliver. But if we truly believe that the officers we send to in-residence graduate programs are shirkers eager to ply their trade—which means that the faculty are not mentors and role models but professional babysitters—then what is the point of having a staff college in the first place?

One indispensable feature of an elite staff college would be academic freedom. Military professionals are often wary of this idea, and not without reason. Armies exist to intimidate or liquidate opponents on the field of battle. Reasonable dissent (calling attention to overlooked details during a commander's conference) has its place, but once orders are issued the soldier is obliged to obey them. All of this of course is quite sensible, but the deeply held suspicion that academic freedom might undermine military discipline or bring disgrace on the profession is founded on a misapprehension.

A good definition of academic freedom is: "The open exchange of ideas that bear on operational matters, without fear of intimidation or reprisal." Academic freedom is not meant to provide cover for the trumpeting of weird notions, rather, it is quite the opposite. The military professor does not traffic in opinion. The sum of his experience, education, professional conscience, and academic position sets him apart from his peers by qualifying him to seek truth, not for personal gain, but for the good of the nation. Every once in a while a military professor might assert an unworthy idea or otherwise draw hasty conclusions that are expressed in a clumsy or abrasive manner. But imperfections of this kind are rare, and placing restrictions on academic freedom because of them would do a great deal of harm. The custodian of academic freedom at our notional staff college would be the president, a person of high intelligence and stout moral courage. The president must have wide experience in both military and academic matters, and approach the demands that each would make on the college without prejudice. The president must demonstrate a keen understanding of how innovative ideas are generated and, correspondingly, have a firm comprehension of the sources of resistance to them.

The location, curriculum, and pedagogy of a reformed staff college are inert if we are not careful as to who teaches at the college and who attends. Faculty should be drawn chiefly from three sources: graduates of the staff college who demonstrate exceptional professional and academic ability, military officers who have served as faculty at the other war colleges and service academies, and civilian professors with distinguished scholarly and teaching credentials. The hiring process should produce faculty members who relish teaching and have proven to be not only efficient custodians of their careers, but have demonstrated an intelligent appetite for studying their profession. Academic promotion will fall in line with the standards that obtain at reputable civilian graduate programs.

We should borrow the practices of the service academies and set age parameters on officers seeking admission. Prospective students should hold the rank of captain or major (pay grade of O-3 or O-4), and have completed a minimum of three assignments. They should not be younger than, say, twenty-seven years of age. The upper limit should be thirty-five years of age—though carefully scrutinized exceptions might be made for officers whose field experience was delayed at the start of their careers because they won a Rhodes or Marshall Scholarship, or otherwise attended an elite graduate program under the sponsorship of the service academies. In considering officers for a place at our joint war college, the admissions process must screen out those officers who have fallen into the narrow groove of their careers, signs of which would include an uncritical faith in received opinion, and the inability to write clearly.

There are plenty of officers who have built successful careers, but for whom reading, say, Edward Gibbon's *The Decline and Fall of the Roman Empire*, would be a huge waste of time—provided that they weren't overawed after perusing the first couple of sentences. In that book there is a vignette (the present writer cannot recall the specific chapter) that justifies why our joint war university should be created to serve officers with demonstrable academic ability, and ultimately, to create an elite corps of senior leaders. A barbarian tribesman, part of a band that sacked a Roman settlement, comes across coins in a leather sack. He throws away the money—an artifact of high civilization— and keeps the bag. In much the same way, officers with little intellectual aptitude attend a war college to collect a credential, something they can use to advance their careers; expecting them to profit intellectually is akin to expecting the Tartar to use the Roman coins to buy a copy of Cicero's philosophical epistles.

In sketching out a first-rate war university, one has to take into account commissioning programs, because it is during one's undergraduate years that intellectual vibrancy takes root. In order to build a measure of coherence between the lieutenant and the prospective student at the war university, consideration should be given to how we train officer candidates. Reserved Officer Training Corps (ROTC) and service academy cadets with superior academic records, particularly in history, philosophy, political science, and English, would be the ideal choices for future assignment to the war university, and beyond that, senior command and staff duty.

To make sure that our intellectual capital is of high quality, the services should award more ROTC scholarships, and the USAF should begin doing so, to cadets who choose the humanities as an academic major. One primary reason that the United States has not in recent times produced a military theorist of the caliber of Clausewitz or Mahan is because we continue to encourage, indeed demand, that officer candidates have a strong grounding in the engineering sciences. That the armed forces need officers with degrees in engineering is of course very true (for reasons so obvious that no discussion of this point is necessary). But we also need officers with degrees in the humanities, given the character of joint war planning in particular.

True enough, planning a military operation often boils down to straightforward questions that bring to mind engineering problems. What weight of effort is required to subdue the enemy? By what means are we to move forces, and are there places of embarkation and debarkation that can support operational timetables? What effects will climate and geography have on a campaign? Is there a suitable volume and quality of logistical and communication assets ready to hand? Even so, this information is by itself of limited value. Unless we have a full understanding of the prospective enemy—how he thinks, what is the likely range of his actions and reactions, what motivates him, what are the sources of friction that might frustrate his war aims—we will sooner or later be faced with unpleasant surprises. Yes, at the end of the day we will prevail on account of our immense material strength and the inexhaustible and intelligent bravery of our forces. But the cost in lives, treasure, and national morale will almost certainly be less if we looked upon the enemy as a human being, rather than as a malleable object or system—which is precisely the foundation upon which joint doctrine is currently built.

John Derbyshire, a connoisseur of technical innovation, perhaps best known for his engagingly written books on mathematics, offers an acute observation on the limits of science that operational planners and senior commanders would do well to ponder. "The arts and humanities are not mere entertainment, to be turned to for relaxation after a busy day of solving differential equations," Derbyshire writes in an essay aptly titled, "Blind Science." Rather, philosophy, history, and English are "templates for living, for governing ourselves and our societies." "Nor can science offer any help with the knottier problems besetting the human race," Derbyshire goes on to say. "It can remedy bad smells, bad pains, and bad roads, but not bad behavior, bad government, or bad ideas."[11]

A humane education sharpens the military mind in ways that engineering cannot. Work in the humanities requires the careful weighing of evidence, the due consideration of competing views, and the clear, efficient expression of complex ideas. These skills can only be acquired by wide reading in the masters of English prose, by extensive and demanding practice, and by studying under the tutelage of highly capable teachers found in our nation's top liberal arts colleges. It is not enough merely to take a class in composition, enroll in historical and philosophical survey courses, and flip through writing handbooks. Of even greater importance is that the humanities enlarge one's inventory of experience. The cultivation of historical-mindedness in particular is vital to commanders and planners. Far better for military officers to possess a discerning grasp of campaigns from the past, which provide a record of things actually accomplished or thwarted, than to put faith in abstractions or mechanistic approaches to waging war.

The humanities, moreover, encourage a spirit of emulation: the motivation to strive after and surpass flesh-and-blood examples of moral excellence. The triumph of armies struggling in the face of what seem to be insurmountable odds can inspire, even as the narrative that explores a commander's costly blunder can encourage readers to recognize the same mistake in themselves, and thus forestall or subdue similar habits of mind. What is more, the officer well-schooled in the humanities (assuming he has taken his studies to heart) is more likely to see the practical and transcendent value of humility, patience, and an intelligent awareness of human fallibility: character traits that make for a staunchly ethical, wise, and perceptive leader.

Nowadays we make great play of the importance of understanding other cultures, but such is impossible without the right frame of reference. One can't claim to understand Germany, Venezuela, or China simply because one has completed a short course in the appropriate language. Ideally, one would master the language as a first step, which was followed by an intensive study of the country's historical, philosophical, and imaginative literature. Reading these works in translation is an acceptable alternative. Having gained a sympathetic understanding of a given country's culture (i.e., its common historical memory), the officer would be qualified to make judgments as to how a nation and its leaders might likely behave, because people within a given culture tend to act in ways that, while not always predictable, fall within a finite compass. All of this, of course, must be built upon a thorough grounding in the history and culture of the United States.

A subtle, but perhaps decisive, advantage of the humanities over the engineering vocations is the way in which graduates of such programs tend to absorb and transmit information. Engineering equations and the characteristics of inanimate materials can be reduced to bullet statements, graphs, and the like, which is perhaps at the root of our promiscuous reliance on PowerPoint™. The engineer inclines toward formulaic solutions and instinctually seeks linear means of understanding and solving problems. While teaching at the Joint Forces Staff College, the present writer was often dispirited by students who preferred a PowerPoint™ presentation to substantial reading, on the grounds that they were "visual learners," trade-school jargon that reflects an impoverished understanding of what constitutes knowledge.

The engineer, moreover, is by definition a specialist; civil engineering differs sharply from astronautical engineering, which bears little in common with mechanical engineering, and so on. It's worth recalling also that the study of engineering is unmoored from disciplines such as history and the philosophy of war that are pivotal to effective war planning. One can buttress an engineering curriculum with core courses, but surveys are nothing more than introductions and are not designed to create a solid understanding of a field of knowledge. And many engineering students are encouraged to approach required courses in "fuzzy" subjects with acquiescence, rather than with a proper appreciation of their worthiness. By contrast, the graduate of a humanities program is much more likely to be a discriminating judge of human affairs in all its variety, and so is especially well suited to war planning. Apologists for the engineering sciences can make no such claim. In a word, the engineer's skill is essential, but not nearly comprehensive enough to meet the demands war planning imposes on staff officers. An operational planning staff needs both types of intellect, but the outlook provided by a humane education is very much muted if our manner of selecting and training staff officers, and our embrace of the system-of-systems approach, are anything to go by.

Assuming that officer-commissioning programs take proper account of the value of a superior undergraduate education, the next administrative challenge centers on our approach to joint assignments in particular. The reigning assumption nowadays seems to be that a joint staff job for midcareer officers is either an interlude between operational assignments, or a pre/postcommand finishing school in which the obligatory, or at least highly desired joint billet, is reflected on an officer's promotion dossier. The joint staff job is also treated as a diplomatic mission of sorts, in which the officer's role, at least implicitly, is that of an advocate who pleads the case for his service. These approaches, while understandable and even beneficial to the military departments, imply that joint staffs ought not to comprise an elite troop.

Much has been said here about rehabilitating our joint education program, but even ambitious reform will be of limited value if promotion and assignment priorities are not recalibrated to take full advantage of the graduates of a reformed joint war college. The United States armed forces have never put much faith in the idea of a general military staff, and this perhaps accounts for the whiff of disesteem that surrounds such duty—particularly when we recall the intellectual feebleness of our current Joint Forces Staff College. Joint staff duty ought to be a career field that is complementary to, and perhaps even separate from, the command track, and distinct from duty within one's service. As mentioned previously, prospective staff officers would have a strong academic record ,and many would possess a degree in the liberal arts. They would devote the first few years of their careers to building expertise in their branch or weapon specialty, and collaterally, acquire a firm understanding of their services' core competencies and doctrine.

To those who have spent their careers within a given service this assertion might seem reckless (it is not unreasonable to expect one's military department to have first claim on one's professional loyalty), but it cannot be emphasized enough that we fight as a joint force, so the joint force perspective must domineer over that of the individual services. Six or seven years spent on active duty within one's service, which is built upon service-specific commissioning programs, should be more than adequate to give the junior officer a mature understanding of the army or the navy. During this time, and throughout their careers, officers should allocate time for professional reading as a means of fortifying this understanding. And it wouldn't be a bad idea if occasionally officers wrote essays or notes for publication that reflect their experiences, something that might improve the quality of military journals.

In fact, there should be an informal encouragement to do these kinds of things. After all, military service is a profession and not merely a job; physicians, lawyers, and other members of the professions advance, in part, by meeting similar expectations. Far better for officers to spend such spare time as they have reading in the literature of their profession, and perhaps making particularly good use of their knowledge by writing for publication in military journals, rather than pursue an online civilian graduate education, or attend night school, the motivation for which is often to fill a square for promotion. Academic study under such conditions is otherwise pretty much worthless. Most of these degree programs are no more rigorous than high school advanced placement courses, and the instructors are often part-timers or adjuncts, whose motivation is chiefly to earn a little extra money. The military profession deserves better.

As far as personnel management goes, the needs of the joint staff would supersede those of the military departments. If an officer earns entry into the joint war college, his service should be obliged to release him. Options for the joint war university graduate would include duty at any of the combatant commands, or service on the joint staff or on the various headquarters staffs of the military departments. Subsequent tours would include command, and perhaps a tour of duty as a professor at the joint war university. Promotion opportunities would also reflect the importance of graduation from the war university. Graduates of the program should be well represented among general officers.

The idea of giving promotion and assignment priority to officers educated to excel at military planning, will doubtless generate a range of responses, from derision at rewarding overly intellectualized officers, to disgust at the possibility of resurrecting "Prussian militarism." But one major fact that our current personnel system overlooks is that ideally, the temperament of command differs from that of the staff officer. Our current means of grooming officers for high command strives after a uniformity of experience and outlook that is controlled by service-specific doctrine and culture.

It is worth recalling that the habits of mind that lead to success in command are often at odds with those needed to produce excellent staff work, operational planning most especially. Officers advance to command by demonstrating the ability to make decisions rapidly in the face of risk, based on judgments anchored in doctrine and experience. A firm grasp of conventional thinking is thus necessary. Hardly less obvious is that the up-and-coming commander builds a reputation by knowing the mind of the boss and acting accordingly. It is also true that command requires building consensus and acknowledging the practicality of compromise—circumstances that cannot help but dampen the impulse for bold, innovative thought. Operations officers and commanders, moreover, are guided by the day-to-day administrative and technical demands of their unit and career specialties; no time can be set aside for speculative ideas. To work outside these channels is to throw a spanner into a hard-working machine. In other words, the best commanders demonstrate, among other virtues, a capacity for absorbing detail, and a keen instinct for acting within the bounds of accepted points of view. Walter Bagehot's characterization of Sir Robert Peel might well be used to describe a successful commander: "a man of common opinions and uncommon abilities."[12]

The assumption nowadays is that staffs should be comprised of specialists who can translate their technical or vocational know-how into effective war plans. There is an essential wisdom to this manner of proceeding, needless to say. But moving from the fast-paced operational world to what should be the exacting intellectual environs of war planning is for many officers not an easy move to make. The ultimate issue should not be, "Do we have the right spectrum of specialists on the staff," though of course this question must be answered satisfactorily, but "Do the specialists demonstrate the intellectual aptitude that staff work demands?" By definition, specialists—the pilot, the logistician, the surface warfare officer—are not trained or encouraged to take a panoramic view of things (which explains in part the complaint of senior leaders that midgrade officers tend not to think strategically). It can be said that military officers, much like legal, medical, and business professionals, know more about less as they advance in rank. One is expected to become a guru in one's career field or branch of service; but the exercise of an encyclopedic knowledge of the employment of artillery or the F-22 means that there is no time to spare for acquiring broader knowledge, or for reflection on the relation between theoretical ideas consecrated by time, and current approaches to waging war.

Of course, there are officers who have succeeded brilliantly at both staff and command. Human beings don't fit neatly into categories; there are plenty of examples that disprove any ironclad assertion that great commanders cannot be great staff officers. Helmuth von Moltke, for example, did not command in the field until he was a sexagenarian general, yet his performance in the Franco-Prussian War is legendary, eclipsed only by the excellence of the German General Staff system that he brought to perfection. The point is that as a general rule our personnel policies dilute the talents of officers who tend toward command and those who incline toward staff work. Commanders inevitably will gain some staff experience in their career, just as the aspiring staff officer will not enroll in the college unless he has served in an operational specialty. But our personnel management policies should be readjusted so that we can more efficiently exploit the abundant intellectual capital of our officers. There ought to be a way, other than command, of attaining senior rank, if only because the complexity of war demands as much. A joint staff college built on sound principles is essential to developing an elite cadre of staff officers. And as subsequent chapters demonstrate, victory in war often hinges on intellectual qualities as much as it does on valor and tenacity.

Chapter 4

From the Ardennes to Dunkirk: France 1940

The U. S. Armed Forces are in danger of becoming thoroughly marinated in the habits of mind that the seventeenth century termed *clerkes* and *mechanicks*. We act as if communications, satellites, precision weaponry, and clever organizational charts brought to life by smooth-functioning organization men, who can be counted on to obstruct the infiltration of unsafe ideas, are the things upon which victory pivots. The true enemy of authentic military reform is not an insufficient volume of modern machinery and fashionable management theories, as transformation literature would have us believe, but our repudiation of the past.

Our faith in technological wizardry aggravates rather than compensates for the imperfections in our strategic thinking. A strategic outlook that is fueled by a missionary's zeal to remake human nature, and that views war as a problem that can be approached in a manner not essentially different from revitalizing a business, is more likely than not to act unwisely, without a due consideration of cultural circumstances and the imponderables of human conduct that determine the fate of armies in war and the quality of the peace that follows. Given these circumstances, it is all the more critical that military officers demonstrate a much wider view of strategy and operations than anything a clever technician or a career-minded administrator can be expected to comprehend. The advice offered to civilian leadership should be the fruit of solid learning and purposive reflection operating on experience. This is an officer's duty.

The "issue of war and battle" turns on moral factors, Basil Liddell Hart tells us. "In the history of war they form the more constant factors, changing only in degree, whereas the physical factors are different in almost every war and every military situation."[1] The truth of what Hart says can be illustrated by even a brief consultation with military history. In this chapter and the two that follow, campaigns from the past are interpreted as an expression of intellectual and moral strength, rather than as contests between machines and bureaucratic arrangements.

The campaigns are drawn from the Second World War for a couple of reasons. First, the impact of a given campaign was hardly ever isolated; studying them shows the relation between strategic and operational planning. Second, World War II gave rise to the phrase (endlessly repeated on television and in popular history) of the "wonder weapon," the trope that encourages us to focus on armaments, rather than on the man wielding them. Before then, at least between the age of Clausewitz and theoreticians such as Jean Colin, the morale of troops and the intellectual ability of commanders were accepted as the engine of victory. The writings of airpower theorists such Giulio Douhet, along with the experiences of the Great War—when morale didn't quite stand up against machine guns, poison gas, tanks, and massed artillery fire, and when gifted generalship was a rarity—began the intellectual shift away from the mind and character of the soldier, to the various sorts of firearms he might employ.

The publication of books by Jane's (*All the World's Fighting Ships* first appeared in 1898, followed by *All the World's Airships* in 1909) adumbrated the eclipse of moral and intellectual mettle as the commonly understood foundation for victory. In the wake of the Second World War, books devoted to the study of weapons were published on a massive scale, an artifact of that conflict unique in the annals of warfare. An example that springs to mind is that the estimable *Ballantine* monographs on the Second World War featured an entire series of (quite good) "Weapons" books. Among the titles are *Waffen SS: The Asphalt Soldiers, Airborne, The German General Staff*, and *Commando*, thus suggesting, inadvertently one supposes, that specialized troops are indistinguishable from grenades and battleships. Perhaps a more fitting series title would have been "Elite Forces." One other concept that became a de facto weapon during that conflict was ideology: national socialism, bolshevism, democracy and, in the case of Japan, nationalism born of a deep-rooted cultural consciousness (the racialist element in National Socialism was a recently manufactured idea). Soldiers and civilians were told that they were servants of these things, even though the reality on the front lines was of a more complex cast.

The tools of war are important, but they are not the primary agents of victory. This chapter and the two that follow call attention to the importance of strategy and the conduct of operations not as abstract ideas, but as expressions of human frailty and resourcefulness shaped by culture. For all the beguiling magnificence of guns and ships and planes, war remains at heart the most intense form of social intercourse. Its outcome turns not on how well a given set of weapons is served, but on the moral and intellectual constitution of soldiers and their commanders.

A few readers may wonder why the campaigns discussed here do not feature the armed forces of the United States. Our involvement in the Second World War has been exhaustively covered not only in books, but also in motion pictures and documentaries. Over the past fifteen years the anniversaries of battles fought by the United States were widely celebrated in the print and news industries, and enshrined in monuments across the country, many of which have been raised very recently. And there are large numbers of Americans who have at least some vicarious experience with the Second World War from veterans who have generously, if at times reluctantly, talked about their experiences. By contrast, the campaigns surveyed in this and the two subsequent chapters are likely to be much less familiar—if not unknown—to American readers, even though they were immensely important at the time and, as I hope to show, remain compellingly relevant to our current circumstances.

Defeat is a wise but stern preceptor. Our inexperience with military catastrophe imposes on us the obligation to learn vicariously from what the Germans, British, and French were taught firsthand and at great cost. Perhaps the ensuing commentary will encourage readers to study in suitable breadth and detail the campaigns discussed here, which are every bit as engrossing as Normandy and Iwo Jima. This book will have succeeded in reaching one of its objectives if readers are convinced, or reminded, that no military campaign can be treated effectively in a few pages. The aim here is not amplitude or comprehensiveness, but rather to highlight, by way of historical analogy, the need for an intellectual transformation in our armed forces—an authentic transformation, not the flimsy item shrouded in gauzy rhetoric, that is peddled today. The industrial and technical might of the United States is likely to remain unchallenged for some time—hardly a bold prediction given our national character, which is defined in large part by our technical resourcefulness fueled by ever-expanding wealth. If we are at risk it is because of the dry rot of technology-worship that, over time, has enervated the intellectual vitality of the officer corps.

It might well be asked what possible use can studies of past campaigns have to the fourth-generation warrior or an avatar of transformation hoopla? To begin with, studying history inoculates the critical intelligence against embracing such fatuities as the belief that our experiences today are uniquely deepening and unprecedented. Reading history ("the essential corrective to all specialization") helps the military officer cultivate disinterestedness and perceptiveness, virtues that do not come naturally to the bureaucratic mind.[2] The study of history is not a recipe for clairvoyance; accident and human agency make certain of that. Only someone who claims to know the mind of God would assert that history allows him to predict with certainty what will happen in some future place and time. What we can accomplish by studying campaigns from the past is to acquire a perceptive understanding of folly and its sources, because many military blunders are at heart the repetition of earlier patterns of faulty thinking. The conscientious historian discerns causation among a welter of facts, many in apparent contradiction to each other, and by doing so illuminates the murky and entangled connections that unite motive, action, and consequence.

In opposition to this, one can easily imagine the despisers of the military intellectual asserting that experience is what matters most (time in the field is more valuable than time in a library), and such advanced education as the up-and-coming officer needs will be provided by the various war colleges. Such an outlook, prescribed by custom and validated by the decisions of promotion boards, is not difficult to rebut. First, experience can only be animated by reflection, otherwise it remains inert. On this point one would do well to recall the observation of Frederick the Great (here paraphrased from memory), "if experience were all that mattered, all my pack mules would be generals." A moment before typing these words, an e-mail reached the present writer seeking candidates for Reserve Officer Training Corps professors. Deployment experience was highly desired, but no mention was made of intellectual ability, which should not be confused with possessing the necessary credential of a master's degree. That an officer might have obtained a degree in, say, management, from an online university or night school, hardly qualifies him to profess on strategy and operational campaigns.

Scarcely less important is that firsthand experience in war, which even a full career is necessarily occasional and narrow; much of one's time is absorbed by the routines of training and administration. The fighter pilot may have 2,000 hours in the cockpit, but what proportion of this time will comprise actual combat, and what does the sum of that experience tell him about the strategic impact or operational demands of maneuvering an army across a vast, trackless desert, or fathoming the motivations and aims of statesmen and commanders in the service of an alien culture? Such a man is a technician; our expectations of his leadership ability must be rather modest. A remarkable deployment record in itself tells us little about an officer's suitability for command and staff work. Without the broader perspective that only reading history provides, the veteran of modern campaigns is likely to exaggerate the significance of his relatively tiny ration of experience, or otherwise distort its value as a guide to higher command in war.

What follows is a survey of the German invasion of France in 1940. The value of studying this campaign or any other of similar prominence is that we see in play factors that shape the outcome of battle, which can never find expression in the pseudoscientific jargon and the mechanical turn of mind that hamstring contemporary joint war planning. It was the insight and intelligent audacity of staff officers, rather than a corporate adherence to a framework or formula, which set the conditions for the *Wehrmacht's* victory. By contrast, the French plan was superficially reasonable, but destitute of an understanding of the enemy's character and motivation and, for that matter, a disinterested apprehension of France's own forces. It was the very sort of document that the system of systems approach might have validated. More broadly, the campaign shows us the subordinate value of equipment and administration. The judgments of commanders and statesmen and the interplay of character and circumstance determined the outcome. The reader is asked to forgive the frequent repetition of the core argument of this book, but the trajectory of contemporary military culture seems to justify doing so: war is above all else a species of human activity, and not a plastic or mechanical craft. It is fully understood not by a proliferation of paradigms and processes or by graphs, pie charts, or organization schemes, but by a proper study of history.

Eight months after the Allies declared war on Hitlerite Germany in September 1939, the battle for France began. It lasted about forty-five days, but was effectively over after two weeks; the remaining time was spent in what many German soldiers came to see as live-fire maneuver exercises. The ease with which the *Wehrmacht* liquidated the French army in the late spring of 1940 suggests that there was something inevitable about the lopsided victory. German propaganda films, set to the music of Wagner and Beethoven, featured columns of panzers on the move, partially obscured by dust clouds (bringing to mind a stampeding herd of buffalo), or fanning out unmolested on the plains between Sedan and Abbeyville, attended by motorcycle-borne couriers that resembled pilot fish accompanying sharks. Widely published photographs from the period reinforce the idea of German invincibility and Allied impotence: roads clogged with refugees and routed columns of French infantry; grinning and smartly turned-out German soldiers sight-seeing in Paris, which the city's defenders abandoned without a fight; the British Expeditionary Force, bedraggled and denuded of its equipment, making its escape from Dunkirk in a motley of naval and civilian craft; Hitler dancing after the armistice ceremony in front of the railway carriage that provided the scene for the French capitulation. It was Marshal Foch's wagon-lit, in which emissaries from Wilhelmite Germany formally accepted defeat twenty-two years earlier. Hitler ordered its transportation from a museum to Compiégne, the place of the signing of the 1918 armistice, and also commanded the razing of the site immediately afterward. In one of those ironies that fate relishes delivering, the railway carriage was moved to Berlin where it was later destroyed in a British air raid.[3] The humiliation that attended the French defeat was sharpened by the fact that only a fraction of the German army was responsible for it. A handful of panzer formations in concert with a few airborne battalions, out of a total combat strength of more than ten dozen divisions, shattered French morale and brought on her collapse.

As is so often the case in military history, victory and defeat ultimately proved to be more apparent than real. Germany's triumph did not advance her strategic interests in the long term, nor did it irreversibly cripple those of the defeated nations—proving as much as anything else that strategy and its lord-in-waiting, armed conflict, are not beholden to scientific analysis. What is more, only a few of Germany's field commanders acquired substantial combat experience upon which the victory hinged. And even the two most celebrated generals, Rommel and Guderian, still had much to learn about the art of handling large mobile formations. The vastness of Russia, aggravated by that country's climatic extremes and the absence of a modern network of roads, would test the resourcefulness of the *Wehrmacht* in ways scarcely imaginable in the summer of 1940.

The establishment of a French government in exile, moreover, would prove to be at least a mildly vexing problem for the Germans, as the stalwart performance of the Free French at Bir Hakeim in 1942 and the work of the resistance movement proved. The failure of the *Wehrmacht* to destroy the British Expeditionary Force meant that, for all the glory and psychic compensations that attended the subjugation of France, the decisive battle that Hitler hoped for had eluded him. The victory justified Hitler's overblown view of his strategic judgment and adumbrated the circumstances that would lead to Germany's collapse in a few years time. Hitler, his intellect befogged by racial romanticism, assumed that the British, close relatives of the Aryan race and more often a rival than an ally of France, would soon come to terms. This of course was reflected in the complete absence of sensible preparations for an invasion of the British Isles, as well as a breezy indifference to the possibility of a military alliance between Great Britain and the United States in the aftermath of the French collapse. To be sure, isolationist sentiments among Americans remained a political force in the wake of Hitler's conquest of Western Europe, but the reelection of

Franklin Roosevelt in November 1940 should have prompted the German High Command to consider more fully the possibility of American involvement. The passage of the Lend-Lease Act in March 1941, as well as Roosevelt's various speeches during the previous year, should have figured largely in German strategic calculations.[4]

Animated by an enveloping bigotry, Hitler could scarcely be expected to comprehend America's latent military might. As soon as national self-interest coincided with her energetic idealism, the United States could only overpower Nazi Germany. Largely untouched demographically by the Great War, America was endowed with huge reserves of athletic young men. The country loved sport and was awash in firearms, hardly surprising given its pioneer traditions and its largely rural character. Its mythic heroes included Paul Bunyan, a symbol of titanic strength, and Daniel Boone, unfailingly self-reliant and a crack shot. Unlike Hitlerite Germany, moreover, American culture would celebrate rather than recoil from employing women in factories, something one might have been able to predict given the passage of the Nineteenth Amendment (1920). A reader of Tocqueville might have reached the same conclusion.[5]

Hitler believed that Dunkirk was the last act of what had turned into a rout. In yet another twist of fate, what was in fact an evacuation pulled off on a wing and prayer was converted into an inspiring testament to British resourcefulness and tenacity. The British Navy remained a force in being, and the battle-tested troops who escaped would form the core of an expanding army. For the Germans, this meant that France would have to be garrisoned; only the southeast portion of the country could safely be entrusted to a collaborationist government, and even that came to an end in November 1942, just at the moment when the war in Russia was taking an ominous turn. That Hitler took the possibility of an invasion seriously finds expression in his decision to move two *Waffen SS* divisions from the Eastern Front to France in the wake of the Dieppe raid, August 1942. The First SS Panzer Division, "Leibstandarte Adolph Hitler," and the Second SS Panzer Division, "Das Reich," were both used in the November 1942 occupation of Vichy France. These divisions were desperately needed on the Eastern Front that autumn; they would be sent back in early 1943 to take part in the counteroffensive that reversed Red Army attempts to eject the *Wehrmacht* from the Donets basin in the aftermath of Stalingrad.[6]

Shaped by his experiences in the trenches of the First World War, and also by the memory of the French triumph at Verdun, Hitler could only believe that the defeat of France meant that Russia (a semi-primitive country beaten by the Germans in the first war, populated by a non-Aryan race, and governed by a bankrupt ideology that, among other things, was given to self-injurious stupidity, exemplified by

Stalin's murder of his most talented commanders in the late thirties) could not possibly stand up to the *Wehrmacht*. And any caution expressed by Hitler's generals in regard to an invasion of Russia would be in his eyes yet one more expression of their contemptible timidity and dislocated strategic perspective.

Impressive though it most certainly was, the German victory over the French in 1940 was by no means predestined. We should examine the German operational plan, not only from its conception to execution, but also the cultural and political circumstances that provided the occasion for it. The French plan deserves similar scrutiny, for it embodies an approach to planning that, while outwardly conforming to what we call today operational art, suffered from the absence of an intelligent understanding of the enemy's mind and character, and an overestimation of the strength of France's strategic position.

May 1940 was the culminating event of an ongoing rivalry that first took the form of armed conflict in 1870, when Napoleon III declared war on Prussia in an effort to stabilize his faltering domestic authority and, collaterally, to keep Prussia from overtaking France as the dominant power in Europe. The chancellor of the North German Confederation, Otto von Bismarck, didn't shrink from fertilizing Napoleon's belligerence because of what he saw as the advantages to be reaped from war. Prussian forces liquidated French military power within a matter of weeks, and shattered not only the army, but the government as well.

France was humiliated by the terms of the surrender. The Germans insisted on a victory march through Paris, and the coronation of King Wilhelm I took place at the Palace of Versailles, the traditional seat of French monarchs. The Germans demanded a punitive war indemnity, and her troops did not leave France's northern provinces until the debt was paid in 1873. Alsace-Lorraine was ceded to the newly formed German Empire. There were additional, unpleasant ramifications. Imperial Russia abrogated the Treaty of Paris (1856), which had restricted her navy to the Black Sea, and from then on became a force to be reckoned with in European affairs.

French resentment percolated over the next forty years, and this, along with a general deterioration in what was at least a superficial European harmony, was brought to maturity in 1914. The foreign policy of France in the decades following her army's mortifying defeat at Sedan was balanced between rehabilitating her strategic position by expanding her colonies outside Europe, and by building alliances (the most significant was the military pact with Russia, signed in 1894) that were designed to hem in German diplomatic and military aggression. The revanchism and truculent self-esteem that informed the strategic priorities of France found expression in the primacy of the offensive in her military doctrine, an idea conspicuous in the writings of Jean Colin (1864–1917)—the book to read is *Transformations of War* (1900)—who believed that the offensive was the only way fully to sustain and exploit high morale. The offensive spirit also very much informs the works of Ferdinand Foch (1851–1929), specifically his book, *On the Principles of War* (1903).

The Great War turned on their head the strategic circumstances that had been in play for nearly half a century. In the aftermath, arrogance remained as conspicuous as ever in France's cultural outlook, but it was no longer fortified by national will or military power. France had recovered her lost territories, but at a horrifying cost in lives, money, and national morale. One million seven hundred thousand French soldiers perished. Many more were wounded, a fact symbolized by the victory parade of 14 July 1919, which was led not by triumphant generals or troops in the sartorial splendor that is often a striking feature of a military parade, but by "three young men, or what remained of them; unspeakably crippled by war, still in uniform, and trundled by their nurses in primitive chariots like the prams of deprived children."

> Immediately behind them came a large contingent of more *grands mutilés*. Officers and men of all ranks mixed together, many already in mufti, they marched—or hobbled—without precedence or any semblance of military order, twelve abreast. Hardly one had not lost an eye or limb, and many bore on their chests France's most coveted decoration, the Médaille Militaire. The totally blind—some accorded the privilege of being ensign-bearers—came led by the one-legged or armless; men with their destroyed faces mercifully hidden behind bandages; men with no hands; men with their complexions still tinted green from the effects of chlorine; men with mad eyes staring out from beneath skullcaps which concealed some appalling head injury.[7]

From this scene can be drawn only one conclusion; the victory had broken not only the spirit of the army, but had undermined the moral constitution of the nation as well. Many of France's brightest young men, who would have provided a vigorous political, intellectual, and military leadership after the war, lost their lives, their health, or their patriotism on the battlefield. A whole generation of young men had effectively been removed from France's demographic makeup. It is a rather uncouth exaggeration that nevertheless conveys a core of truth, but France in 1919 was a country of widows, invalids, gestating spinsters, and old men. Hardly less important is that the enormous physical destruction of the war (much of the French countryside was ruined) meant that France would be economically supine for several years. There would be little money or political will for military reform, a circumstance that was aggravated by the long-standing French contempt for large, professional armies and militarization in general.

In the two decades that intervened between the world wars, France was beset by a multitude of problems that broadened the chasm between the country's exalted view of herself and her actual stature among European powers. On the domestic front, the only stable French government that might have been recognized before the end of the Second World War by a young adult born in, say, the southeastern part of the country in 1913, might very well have been the one imposed by the Nazis, on the extremely modest grounds that the early Vichy years (1940–1942) represented the loss of national self-determination mollified by political certainty, of however a distasteful character, and what must have seemed like the merciful absence of an occupying force. Political scandals were rife during the interwar years, which the French seemed rather to enjoy. Labor disputes were frequent and often found expression in violence.

Undermining further the national security of France was a robust antimilitarism that happened to coincide with Hitler's rise to power. The French situation dramatically illustrates the importance of culture to war planners. The contemporary literature and philosophy of France reflected far more clearly her military feebleness than any table of equipment and organization.

The postmodern existentialism propagated by Jean-Paul Sartre and his epigones during the interwar years, along with the emergence of surrealism, reflected a national unwillingness to confront an unpleasant recent past, a dispiriting present, and an uncertain future. The best French writers turned out works of fiction that dwelled on the brutality of trench warfare, and turned a skeptical eye on the political and moral justifications for war.

In 1935 there appeared a play written by Jean Giraudoux titled *La Guerre De Troie N'Aura Pas Lieu* (*The Trojan War Will Not Take Place*), which, in its ironical antiwar temperament, bears a remote resemblance to Aristophanes's *Lysistrata*. Twice wounded during the Great War, Giraudoux (1882–1944) served in France and the Dardanelles and was awarded the Legion of Honor. For a brief period after the war he resumed his embryonic career as a diplomat, which included duty as Embassy secretary in Berlin during the early 1920s, but he soon realized that his intellectual gifts were better suited to literary rather than bureaucratic pursuits. Giraudoux's fiction and drama of the period were widely esteemed. By the mid 1930s he was considered "the major French playwright of the twentieth century, and no one else up to that time had succeeded in capturing the attention of both the critics and the public."[8]

There are no battle scenes in *The Trojan War Will Not Take Place*; rather, the play takes a satiric view of the moral fecklessness of political leaders, who declaim on the inevitability of war, even as circumstances argue otherwise. In the following passage, the leader of the Greek federation, Ulysses, explains to Hector—the head of the Trojan delegation, who, having seen enough of war, is committed to avoiding it—that diplomacy is nothing more than preening: a psychic poultice that does nothing more than soothe the consciences and massage the vanity of the heads of state while disguising the ambition, rivalries, and atavisms of the nations they represent:

> You are young, Hector! It's usual on the eve of every war, for the two leaders of the peoples concerned to meet privately at some innocent village, in a terrace in a garden overlooking a lake. And they decide that war is the world's worst scourge.... They study one another. They look into each other's eyes. And, warmed by the sun and mellowed by the claret, they can't find anything in the other man's face to justify hatred, nothing, indeed, which doesn't inspire human affection.... They really are exuding peace, and the world's desire for peace. And when their meeting is over, they shake hands in the most brotherly fashion, and turn to smile and wave as they drive away. And the next day war breaks out. And so it is with us at this moment. Our peoples, who have drawn aside, saying nothing while we have this interview, are not expecting us to win a victory over the inevitable. They have merely given us full powers, isolated here together, to stand above the catastrophe and taste the essential brotherhood of enemies. Taste it. It's a rare dish. Savour it. But that is all. One of the privileges of the great is to witness catastrophes from the terrace.[9]

The cynical outlook of *The Trojan War Will Not Take Place* foreshadows the timid reaction of France's military and civilian authorities to German aggression in the years leading up to the Second World War and, indeed, after the panzers crossed the frontier in 1940.

The pessimism that suffuses Giraudoux's play also finds expression in Henri Barbusse's novel, *Le Feu* (the English-language version is titled *Under Fire*), which first appeared in 1916, and is based on the author's experiences at the front the year before. *Le Feu* remained popular during the interwar years, and introduced the revolting circumstances of trench warfare to a generation of French readers too young to have endured it firsthand. But the novel does not celebrate the pornography of violence; rather, Barbusse argues a political point, as the following passage makes clear. In the aftermath of a battle the narrator explains to fellow soldiers that the war is being fought not in defense of liberty, but on behalf of "financiers, the great and small wheeler-dealers, encased in their banks and their houses, [who] live by war and live in peace during war."

> And even when they say that they do not want war these people do all they can to perpetuate it. They feed national pride and love of supremacy

through force. "We alone," they say, each behind their barrier, "we alone possess courage, loyalty, talent and good taste!" They make something like a consuming sickness out of the greatness and richness of a country. . . . They are your enemies as much as today the German soldiers who are lying here among you and who are nothing but poor dupes. . . .[10]

These words were originally written two years after the German defeat on the Marne imposed the stalemate that led to trench warfare, but throughout the 1920s and 1930s they came to reflect and endorse a clutch of popular resentments and will-o'-the-wisp sentiments that infected French diplomacy and military strategy.

Henri Barbusse died in Moscow (he was an avowed communist) on 30 August 1935, six months after Hitler publicly repudiated the restrictions of the Versailles Treaty, and about six months before the *Wehrmacht* occupied the Rhineland. Nearly a third of a million people lined the streets of Paris to watch Barbusse's funeral procession—itself an oblique indicator of French moral exhaustion.[11]

The French lionized antiwar writers, but in Germany the author whose international reputation was greater than any of the authors living west of the Rhine was treated as a de facto criminal. Published in 1929, *All Quiet on the Western Front* sold more than one million copies in Germany (sales of the book around the world in various translations were of a similar magnitude), but the Nazis banned Erich Remarque's novel shortly after they seized power. In 1933 Hitler's propaganda minister, Josef Goebbels, made a public display of repudiating what the Nazis saw as the unheroic casting of *All Quiet on the Western Front*, burning copies of the book and the Oscar-winning film based on it in front of the Berlin Opera House—a flamboyant expression of resurging German aggressiveness that Western military and civilian authorities should have paid closer attention to.[12]

By contrast, Ernst Jünger's memoir of service in the Kaiser's army during the Great War, *Storm of Steel* (1920), enjoyed official endorsement and was a particular favorite of Hitler. Unlike the contemporary recollections of French and British authors, Jünger's book is defiantly celebratory, even as it does not shrink from portraying the squalor and savagery of war. In the following passage Jünger expresses a point of view that would simmer throughout Germany in the years leading up to the Second World War.

> Hardened as scarcely another generation ever was in fire and flame, we could go into life as though from an anvil; into friendship, love, politics, professions, into all that destiny had in store. It is not every generation that is so favoured. And if it be objected that we belong to a time of crude force our answer is: We stood with our feet in mud and blood, yet our

faces were turned to things of exalted worth. And not one of that countless number who fell in our attacks fell for nothing. Each one fulfilled his own resolve. . . . When once it is no longer possible to understand how a man gives his life for his country—and the time will come—then all is over with that faith also, and the idea of the Fatherland is dead; and then, perhaps, we shall be envied, as we envy the saints their inward and irresistible strength. . . . It was our luck to live in the invisible rays of a feeling that filled the heart, and of this inestimable treasure we can never be deprived.[13]

Readers will not find similar ideas expressed with such intensity in the vast amount of war literature produced in France or elsewhere in the West; certainly not in the fiction of Hemingway, the memoirs of Graves, or the poetry of Sassoon. The closest one that comes to Jünger's work is *Sagittarius Rising*; but here the joy of life on display in Cecil Lewis's memoir comes in spite of, and does not spring from, war's unrelenting ghastliness, and Lewis's perspective is self-consciously private, rather than nationalistic. The literature produced by writers from the victorious nations more or less agreed that the Great War shattered permanently the foundations of a civilization. In contrast, Jünger expressed a point of view that looked to the future as a time of regeneration, followed by the reconstitution of Germany's rightful place as a world power.

The interwar period was hardly more encouraging for France on the diplomatic front. Given French vanity and the enormous bloodletting of the First World War, it is hardly surprising that France drifted into a complacent stupor. For starters, she assumed that in any future conflict the British and Americans would rush to her side; how could it be otherwise given that France was the nurturing mother of Western European civilization? And in any case, France had pretty much defeated Germany single-handedly; on this count history could only repeat itself. As for the material cost of the war, Germany owed France; she and her allies would see to it that reparations were made good—to the penny—even though the draconian terms of the Versailles Treaty were designed to make sure that such was an impossibility.[14]

The Versailles treaty was meant to safeguard French security, but the effect was quite the opposite. The accession of German territory added neither to the sum of her industrial capacity, nor subtracted from that of Germany. Its only effect was to spur Germany's native technical resourcefulness and to provide a compelling motivation for settling the score with France. France's most important ally during the Great War was Russia: she had provided a second front that diluted Germany's offensive power at a moment when France could hardly have parried it at full strength. But Russia's embrace of a political system that the French governing classes looked upon with horror, impelled France to turn elsewhere for an eastern counterweight to Germany. The result was a collection of diplomatic agreements with the small, largely destitute countries spawned by the collapse of the Austro-Hungarian Empire. As "the aftermath of Munich was to prove," writes Alistair Horne, these countries "would just as soon cut each other's throats as form a coherent front against Hitler's Germany." Should France and Germany wage war on each other, France "would in all probability have to rush to the aid of her weaker allies." And this is indeed what happened. Thus, a combination of diplomatic wishful thinking, national arrogance, and strategic myopia set the conditions for France's defeat before the first shots were fired in 1940.[15]

The first trial of moral strength came in March 1936 when Hitler's armies reoccupied the Rhineland. They might easily have been expelled, for at that moment the German Army was unblooded, inadequately equipped, and quite small: only three infantry battalions were used in the reoccupation. The Germans had very few tanks, none of which were used in the Rhineland operation (the first panzer division had been established only five months before), and such tanks as Germany possessed were poorly armored and armed, and did not carry radio sets. The *Luftwaffe* was still in its infancy in 1936. Equipped with only a handful of fighters and trimotor transports, the German air force could be of no practical use to the army. A pair of French divisions under effective command and backed by a suitable reserve might easily have overwhelmed the mouse-sized German force—an event that likely would have wrecked Germany's territorial ambitions and, as likely as not, irreversibly weakened Hitler's political authority.

The French armed forces, however, were not up to the task. As the small *Wehrmacht* detachment began occupying the Rhineland, the French Commander in Chief, General Gamelin, informed the government that while the army was willing to attack, it was far too small to confront the Hitlerite aggressors (an assertion contrary to fact, as Gamelin had to have known), and that mass conscription was needed to flesh out the ranks. With a parliamentary election only weeks away, the French government looked upon a military draft as political self-immolation. So France stood by while Germany moved in and began building the West Wall, thus allowing the *Wehrmacht* to mass their armies in the east, while leaving only a small covering force in the west.

As is so often the case in affairs of state, France's diplomatic actions were shaped by domestic political calculations, which also happened to coincide with the predilections of the French High Command. Whatever the transitory satisfactions that might have come from acquiescing to Hitler's belligerence, France's strategic position soon grew even more precarious after the German Army occupied the Rhineland. Belgium withdrew from her alliance with France and declared herself neutral, thus making a hash of the Maginot Line and the strategic calculations that were built upon it. French authorities had refrained from extending the Maginot Line to the sea, partly out of fiscal restraints, but also for diplomatic reasons. Were Germany to attack Belgium, the Maginot Line would serve as an anvil against which the German army could hammer the Belgian forces, while the French army waited to repel an attack on her own territory. At least, this is how the Belgian government would have looked upon a French fortification that stretched the length of her border. With the Belgian Army no longer willing to act as a breakwater, and in the absence of defensive fortifications, the French army would have no choice but to conduct a fluid battle with German spearheads driving across the Flemish plain, something her armies were neither equipped nor trained to execute.[16]

In 1940 the French Army was hardly ready to conduct any sort of modern offensive campaign, and this can be explained by decisions taken over the preceding century. It would not be unreasonable to argue that for the previous millennium the Gallic peoples had played the role of Europe's chief troublemakers. But from the defeat of Napoleon I up until the First World War, military service among the French was generally disesteemed, even though there was a patina of glamour attached to the impeccably turned out officer, particularly those serving in the cavalry arm. The military profession was shunned by the middle classes. Universal conscription was politically untenable, and in any case the army (not unlike ours today) placed great faith in a force comprising the long-serving professional. The French army had no military staff system worthy of the name. Her military academies offered instruction of indifferent quality. The French government provided funding for attendance at military academies, but remuneration and living conditions among the ranks remained deplorably low. France was often at war during the nineteenth century, and in fact had extended her empire in Africa. In Europe her armies had defeated the Austrians (1859) and the Russians (1854). But these achievements were based not on military excellence, but on the dilapidated state of the forces the French army defeated. French soldiers were brave, and by the end of the nineteenth century, reasonably well equipped, but the brain of her army was vitiated by the culture that produced it.[17]

It is hard not to admire the élan that obtained among French soldiers at the beginning of the twentieth century. That an entrenched enemy might look with apprehension on a bayonet or cavalry charge reflected the valor of the French soldier, but it tells us nothing about the underlying moral and intellectual vitality of the armed forces. The Dreyfus Affair, which transpired over the last decade of the nineteenth century and the first of the twentieth, reflected the tensions of the society at large: republican versus royalist; those who feared and loathed the German versus the international socialist; the middle classes versus the intelligentsia. To be sure, a measure of glory attended military service in 1914, but it was of the kind expressed in Kipling's poem, *Tommy*.

> For it's Tommy this, an' Tommy that, an'
>   "Chuck him out, the brute!"
> But it's "Saviour of 'is country," when the
>   guns begin to shoot;
> Yes it's Tommy this, an' Tommy that, an'
>   anything you please;
> But Tommy ain't a bloomin' fool—you bet
>   That Tommy sees![18]

The sentiment exalting military service that flourished at the beginning of the Great War, as with all emotion divorced from reason, evanesced as the cascade of battlefield casualties touched just about every French family. By 1917 national morale had eroded in ways that would resonate for many years. "Civilian discontent fed military discontent," observes John Keegan in his appropriately titled chapter, "The Breaking of Armies." "The soldiers' anxieties for their families were reinforced by the worries of wives and parents for husbands and sons at the front." The mutinies within the French ranks were quelled by force. Thousands of soldiers faced court-martial; hundreds were sentenced to death; of these, about four-dozen entered eternity not by the hand of the enemy, but by firing squad, while other convictions resulted in life imprisonment. That these numbers seem absurdly small by the lights of wartime losses is yet another illustration of the terminally insidious impact on French society of the Great War.[19]

Things hardly improved in France during the postwar years. The military profession returned to the low status that it had labored under before August 1914. Thanks largely to the terms of Versailles—envenomed by France's visceral urge to punish Germany rather than merely to pacify her—the British, the Americans, and others began to view the French army as a threat to peace rather than its guarantor: hardly a surprise given the cult of disarmament that blossomed in the aftermath of the Great War.

The intellectual culture of the army, never lively to begin with, degenerated further. The defensive-mindedness that took hold of the army in the war's aftermath was consecrated by General Philippe Pétain (1856–1951), hero of Verdun and commander in chief of the

French Army until 1931. Before the Great War, Pétain was out of step with the likes of Marshal Foch and French military culture in general, insofar as he was a proponent of the Clausewitzian idea of defense being the stronger form of war. As Inspector General of the Army during the 1920s, Pétain's advocacy of the continuous front—a faith untouched by the developments of mechanization and the writings of British theorists such as Basil Liddell Hart—squared neatly with the mood of the times. Proposals for an offensive strategy that pivoted on mechanized forces, with all their attendant expense, were not much heard. One exception deserves comment. In 1934 (the war minister at the time was seventy-eight-year-old General Pétain) a slim volume appeared that contested orthodox thinking. *Vers l' Armée le Metier* was written by Captain Charles de Gaulle, an instructor at the military college of St. Cyr. In the year of Pearl Harbor an English translation was published in the United States under the title, *The Army of the Future.*

In *The Army of the Future* de Gaulle asserted that France's vulnerable strategic position, along with the improvident temper of her citizenry, required that she either give up her sovereignty, or develop a robust capacity for offensive action. "By virtue of her physical and mental make-up," de Gaulle wrote, "France must either be well-armed or not armed at all." Assuming the former, France must rely not on a conscript army, but on a small cadre of long-serving professionals, trained and equipped to fight a mechanized war. An armored force suitable to France's strategic needs would comprise some hundred thousand regular troops, and be arranged into six tank divisions and a light division backed up by armor and mobile artillery. Such a force would operate as a strategic sledgehammer, ready to crush an unforeseen attack by the Germans or, conversely, to reinforce a successful defense—pursuit often being a risky enterprise, but one that always promises a large payoff. De Gaulle did not originate the concept of using massed formations of armor and motorized infantry—the British and, later the Germans, were pioneers in theory and practice—but he was the only French officer to argue with intelligent conviction for the application of these ideas.[20]

The relation of truth to error is always one to many. The most enduring works contain a multitude of inconsistencies and imperfectly conceived assertions, especially when they attempt to apprehend the future, and de Gaulle's book was no exception to this. There can be little doubt that de Gaulle's notional six divisions would have dealt a crushing blow to the German spearheads on the Meuse were they to be deployed in good time. It is also possible, indeed likely, that they would have been trapped in the northeast pocket in an attempt to stop German formations advancing across Belgium, because de Gaulle assumed that a mobile German force would strike where the terrain was most favorable to mechanized warfare. Also possible is that the French armored formations would have been smashed in their assembly areas by the *Luftwaffe*—German pilots who had little to fear from the French air force given its doctrinal and material wretchedness at the time.

But even superior equipment placed in the service of a doctrine that conformed to de Gaulle's ideas would not have saved France. At the end of the day the defeatist temper of the French probably would have yielded to Hitler's lust for revenge. Hitler almost certainly would have continued the war even in the face of a sharp operational setback, as the ambitious offensives in Russia after the failure of Operation *Barbarossa* demonstrate. It is also likely that a successful French counteroffensive on the Meuse might have spurred German engineers to field a better tank (French medium and heavy models were superior to anything fielded by the *Wehrmacht*), and this ultimately might have made Germany harder to defeat. On the other hand, had the French shattered Germany's elite mechanized formations, Hitler's generals might have revolted against the Nazis. We cannot know. The point is that the notional impact of de Gaulle's praiseworthy ideas cannot be divorced from the broader military and social context that would bring them to life. It is a truism that ideas have consequences, but military history tells us that at least some of those consequences can never be foretold, and this fact should inculcate a measure of modesty and skepticism in anyone who professes to declaim on wars of the future.[21]

De Gaulle's book is worth reading not only because of its value as an expression of contemporary debates among the French as to how best to guarantee that country's security, but also because it sheds light on our own ideas on transforming the military. Nowadays we tend to believe that the challenges we face are without precedent, but to read de Gaulle is to clearly see the untaught hubris not so much lurking beneath, as gamboling across our transformation literature. In *The Army of the Future* de Gaulle argues that man is beholden to technology in ways that previous generations could not possibly have imagined—a point of view that saturates our contemporary literature on transformation. "Helpful at all times, at present the machine controls our destiny," de Gaulle declares, a remark that aptly captures the spirit of our own outlook on defense matters. De Gaulle is hardly alone in making this point. *The Revolt of the Masses,* by José Ortega y Gasset, appeared a few years earlier and considers rather pessimistically, but no less prophetically, the social and political impact of the machine. But unlike Gasset, de Gaulle says nothing about the philosophical implications of technological advancement, and instead beseeches the French military establishment to embrace technology rather than to place its faith in large numbers of rifle-carrying conscripts.[22]

De Gaulle's advocacy of a lean and technically up-to-date expeditionary force mirrors our own.

> From the day upon which a force shall be created of men of our own country, who are professional soldiers and, in consequence, men prepared to go to distant campaigns . . . and from the day upon which, from time to time, we can parade some of our well-trained troops in carefully selected regions, from that day we shall be sufficiently guarded against danger to render it immediately less probable.[23]

This passage, suitably clothed in the latest buzzwords, would fit comfortably within our own strategic documents.

De Gaulle also makes great play of the idea that a few troops equipped with modern weapons can employ the firepower delivered by thousands during the Great War. Sharing much in common with our outlook, he assumes that wars of attrition are a thing of the past. Total war, in which large numbers of soldiers confront each other in a struggle to the death, has been superseded by wars for limited objectives. "There is a grim relationship between the properties of speed, power and concentration which modern weapons confer upon a well-trained military elite," de Gaulle argues, "and the tendency of nations to limit the objects of dispute in order to be able to seize them as rapidly as possible and at the least possible cost." A decade after these words were published, an army largely made up of conscripts, employing masses of tanks, airplanes, and artillery, supported by an enormous armada of transport ships, would liberate France from an occupying force that was just one part of an army of many millions fighting on two other fronts. De Gaulle's force of 100,000—no matter how technically proficient—would have been much too small to eject the Germans from North Africa, a strategic sideshow.[24]

Not all of what de Gaulle says reflects an uninhibited faith in technology. Apart from his persuasive ideas on using armor in massed formations, he makes a case for education that remains as true in our day as it was in his own. "To exercise imagination, judgment, and decision, not in a certain direction, but for their own sake and with no other aim than to make them strong and free, will be the philosophy of the training of leaders," de Gaulle says.

> Power of mind implies a versatility that does not obtain through exclusive practice of one's profession, for the same reason that one finds it difficult to entertain oneself in the bosom of one's family. The school of leadership is therefore general culture. Through it the mind learns to act in orderly fashion, to distinguish the essential from the trivial, to recognize developments and causes of interference, in short, to educate itself to a level where the whole can be appreciated without prejudice to the shades of difference within it. There has been no illustrious captain who

did not possess taste and a feeling for the heritage of the human mind. At the root of Alexander's victories one will always find Aristotle.[25]

Here, de Gaulle argues that technology and training are of dubious worth if not controlled by a lively critical intelligence, an acknowledgment of a time-tested verity that we have unwisely chosen to ignore, to the extent that we possess the least understanding of it.

The Army of the Future did not change many minds in high places, though de Gaulle's writings did catch the eye of Paul Reynaud. Successful as a practitioner of law, Reynaud entered the French Parliament in 1914—note the year—at the age of thirty-six. His public life would culminate when he became prime minister only weeks before Hitler unleashed his panzers, replacing Edouard Daladier, who along with Neville Chamberlain, yielded to Hitler's demands at Munich. Reynaud promptly appointed Colonel de Gaulle as Undersecretary of the Ministry of National Defense, but by then the cause that both men advocated was long lost.

On 15 March 1935, Reynaud tried to advance de Gaulle's ideas during a debate in the Chamber of Deputies, but he was rebuffed by Petain's successor at the Ministry of War, General Joseph-Léon-Marie Maurin. "How can anyone believe that we are still thinking of the offensive when we have spent so many billions to establish a fortified frontier!" Maurin declared, referring here to the Maginot Line. "Should we be mad enough to advance beyond the barrier—on I don't know what sort of adventure?"[26]

The efforts of de Gaulle and a few others notwithstanding, there was no sustained infusion of intellectual resourcefulness to counterbalance the enshrinement of narrow-mindedness born of the Great War holocaust. How could it be otherwise? The flower of French youth was eradicated between 1914 and 1918. And a young man born after, say, 1905 could only look upon the prospects of military service as a grim one indeed on the grounds of remuneration, status, and quality of life in general. The elderly officers who shaped French military doctrine between the world wars, their judgment calcified by their experiences in the Great War, and hindered further by the pacifistic temper of the nation at large, shrank from the obligations of command.[27]

The recoiling of French military thought upon a rather straitened understanding of the Clausewitzian idea of defense as the stronger form of war, informed its approach not only to the conduct of operations, but also to mechanization, proving that there are few things more self-injurious than a new weapon manipulated by timid souls and narrow-minded intellects. The tank was seen not as a breaker of armies if employed properly and in large formations, but as a subordinate ally of the rifleman. Thus the tanks that entered production in the 1930s (in particular the thirty-two-ton Char B) were heavy, slow, prodigious consumers of fuel, and not distinguished by automotive reliability. It is not that the French didn't entertain sensible ideas. In 1934 the French fielded a combined-arms formation, the *Division Légère Mécanique* (DLM), which featured motorized reconnaissance, infantry, and artillery units, and a pair of tank regiments comprising light and medium tanks, which at the time put them ahead of the *Wehrmacht*. Unfortunately, for France's strategic well-being, the army eventually abandoned this type of organization, because the infantry division was thought to be the decisive formation, along with the concentration of defensive fire exemplified by the Maginot fortifications. The DLM was expected to function not as a war-winning weapon, but as a tactical reserve that would repel any unforeseen breakthrough, and thereafter occupy a holding position. Otherwise the armor could be parceled out, in penny packets, as circumstance required. Put another way, the DLM was dismissed as an extravagance that an increasingly dogmatic and conceited army establishment saw no point in expanding. As war with Germany grew increasingly likely in 1938, the French army created the *Divisions Cuirassée* (DLC), which concentrated the heavy tanks in discrete formations, supplemented by light tanks and motorized contingents of infantry and artillery. But this was a vain effort insofar as the defensive mentality had long since infiltrated French commanders, who remained untaught in how to employ the DLC in the teeth of a resourceful and determined enemy.[28]

France's inability to parry a German strategic offensive is foretold in a book published more than a decade before the event. In 1928 Basil Liddell Hart observed that "the French have created an army which forms a large and powerful, but slow-moving, steamroller of fire which is designed to push back gradually, as in 1918, any similar army which is aligned against it." By contrast, the victim of the French army's heroic stamina during the Great War—and the only nation that could possibly pose a mortal threat to France—had exploited fully the opportunity that a chastening defeat always provides.[29]

That so many fine books have been written about the modernizing of the German Army from the mid-nineteenth century onward means that little needs to be said about the enduring excellence of the German General Staff and the immense influence it exerted on armies around the world. A rather backhanded recognition of the General Staff's achievement was that the Versailles Treaty mandated its abolition, along with the closing of the War Academy, in 1919. But declaring the illegality of a cultural institution is often quite a different thing from realizing its extinction.

The resurrection of the German Army, its manifold achievements defaced by the toxic political machinery it ultimately served, is a story of ideas brought to fruition by individual soldiers possessed of a lively intellect, fortified by immense energy and determination. Their achievement is all the more remarkable given Germany's supine condition in the 1920s. Versailles further stressed an economy much enfeebled by war. Millions were starving. On her western border stood an enemy zealously devoted to punishing Germany. To the east, revolution had installed an ideologically antagonistic regime, buffered by a Poland that was hardly less wary of German intentions, even though militarily impotent. German officers had neither time nor patience to traffic in buzzwords or public relations flapdoodle about "fathering" ephemeral administrative trifles. These men studied their enemies (their writings as well as their operations) in a disinterested and critical way. The Allies were victorious in the Great War, but did they take full advantage of the pioneering ideas, such as employing the tank in massed formations, that were validated by their triumphs? The decisions of the Allies during the interwar years suggested that they did not, thus proving that victory too often spawns complacency, which in turn beckons disaster.

The enterprising officers who rebuilt the German military were not always esteemed by their peers. Friction—most often a self-imposed liability—hobbled the German armed forces in much the same way as it does other military establishments. The innovators fought their share of battles against conventional thinkers within the ranks, at considerable professional risk, and they might easily have been crushed. Perhaps in the interests of peace the quashing of the reformers might have been a good thing. It is, however, a species of historical unseemliness (and pointless besides) to judge the past strictly on the basis of hindsight. History does not deliver up tidy lessons that can be neatly applied to the present. What we can learn from the German army of the interwar years is that any corporate approach to military transformation is best greeted with intelligent skepticism, and that sympathetic attention is owed to dissenting views that are vigorously expressed and closely argued. The outcome of war pivots on ideas as well as on the moral and intellectual constitution of the officers who act on them. Beguiling administrative arrangements, or the latest gadgets, cannot in themselves guarantee victory.

General Hans von Seeckt, the commander in chief, inherited a German army emasculated by Versailles. By the lights of our contemporary understanding of generalship, von Seeckt must appear as an implausible military leader, in some ways a polar opposite of what promotion boards look for nowadays. He never commanded in the field. During the Great War he served as chief of staff to General Mackenson in the east, an experience that instructed him on the great possibilities of maneuver warfare. Pensive, undemonstrative, beholden to no dogma, and a trifle ignorant in regard to purely technical matters, Seeckt was nevertheless a master of strategy and his mind was well trained to exploit his experiences in conducting fluid operations of the kind that propelled the Kaiser's army into the heart of the Ukraine.

Versailles restricted the German army to a ration-strength of 100,000 troops, but this ultimately served to unleash Seeckt's intellectual resourcefulness and his keen aptitude for innovation. In complying with the terms of Versailles, Seeckt made sure the greatly diminished ranks were top-heavy in officers (particularly General Staff members) and noncommissioned officers of high ability. What he thus accomplished was to rescue the brain of the army, even as the corpus, battered by war and demoralized and corrupted by the aftermath (one thinks of the thuggish *Freikorps*), was interred by Germany's enemies. Seeckt established an embryonic *Luftwaffe* by setting up liaisons between the armed forces and the newly hatched Civil Aviation Department. Pioneering work was done adapting gliders to military operations. Exploiting a recently signed Treaty of Rapallo, Seeckt sent officers to Russia, beyond the vindictive eye of the Allies, to experiment with tanks, artillery, and other weaponry. Meanwhile, like-minded officers in the navy (e.g., Admirals Zenker, Raeder, and Wegener) were hard at work. The severe limitations on warship construction merely tapped the German love of technical innovation. Submarine technology was furtively advanced in collaboration with a Dutch firm. The German Navy built smaller vessels that were more lethal than those floated by other navies of comparable, or even much larger displacement, if only on account of their speed.

Firing the intellectual fermentation of the German Armed Forces at this time was an understanding that the Allies never completely grasped. For the victors, the Great War was an exhausting, fierce, and, it was hoped, unique experience; surely, the multitude of treaties, leagues, and disarmament hoopla would see to that. The German military leadership, however, didn't quite see things in the same way. Versailles was a mortifying setback; it would only be a matter of time before the German army would be called upon to set matters right. It was the desire of Seeckt and his disciples to make sure the armed forces were ready to fulfill what many Germans believed was their destiny: hegemony over Europe.[30]

Hitler's rise to power in 1933, for all the suffering and misery it brought to untold millions of Europeans over the six decades that followed, was a boon to forward-looking commanders. Hitler needed an effective army, and by temperament and experience he was bound to have little truck with conventional thinkers. Unlike many French generals, Hitler had no love for the horse; to him, defensive warfare was a catalyst for trench fighting of the kind that could lead to nothing but a collapse similar to that of 1918. Hitler's nefarious political aims thus happened to coincide with ideas generated by bold thinkers such as Heinz Guderian.

Guderian's career exemplifies military innovation in its most intelligent and ambitious form. Guderian was endowed with a questing and incisive mind that often found expression in an irritable impatience with platitudinous ideas and unexamined assumptions, an outlook which Prussian military culture actually encouraged. Officer candidates, for example, "were schooled to acknowledging ultimate authority, but only after argument had been exhausted." The diary of twenty-one-year-old *Leutnant* Guderian bears witness to his appetite for military history, most especially his aptitude for memorizing passages from masters of the art. His lively intellect and strong-mindedness were mutually invigorating traits, which left him with an unsentimental view of his fellow infantry officers, whom he thought of as "insufficiently interested in their profession." Not surprisingly, *Leutnant* Guderian was occasionally estranged from even his closest comrades. "They have lost my respect," he wrote in 1908. "They accuse me of being an introvert . . . but to run with the mob is nothing to be proud of."[31]

Guderian's combat experience in World War I—his work in signals, his instinctual revulsion at static warfare's carnage and the panic induced by Allied tank attacks—gave scope and purpose to his intellectual vitality. In a manner that recalls successful innovators across time, he didn't invent the idea of tank operations so much as adapt the concept to the German army's strategic and economic circumstances. Fluent in English and French, Guderian studied, and in a few instances translated into German, the theoretical works of Hart, Fuller, Martel, and others. His idea was that tank forces, supported by other arms, would surge through enemy defenses on a narrow front and drive into enemy territory in order to attain a strategic objective—a revolutionary idea for its time that was greeted with skepticism and hostility by many of Guderian's contemporaries.[32]

During the interwar years Guderian served as an instructor of transport troops, and he exploited fully the opportunities that faculty duty provided. Confronted initially by skepticism, he refined his ideas along with his skill at presenting them and converted the doubters (who most often were his brightest students) into proselytes. During this time Guderian also published widely in military journals, expounding his ideas to a professional readership. His articles "crystalised his thoughts and his style," notes Kenneth Macksey, and earned him "a reputation for clear exposition on controversial matters." In fact, Guderian's immensely influential book, *Achtung!-Panzer!* (1937), was derived from his academic lectures and previously published essays.[33]

There were other innovators, and many of Germany's most effective commanders (Erich von Manstein, Erhard Raus, Walter Model, Kurt Student, Adolph Galland) were still relatively junior in rank when war was declared. But their achievements reflect in some degree the new ideas that, if not quite given free play in the 1920s and 1930s, shaped the character of operational thought during the Second World War.

Because the means of war are force and counterforce (war is essentially a large-scale duel, a wrestling match, as Clausewitz put the matter), and also because weight-of-effort bears conspicuously on operational planning, it is not unreasonable to begin by surveying the order of battle of the French and German forces in May 1940, even as we acknowledge that this war was at heart one of ideas. Both sides fielded about 135 divisions. The Germans had greater numbers of aircraft, which were of high quality, and better-trained pilots. The Allies held the advantage in quality and quantity of tanks, but German tank crews and commanders were much more efficient, and also had the benefit of recent combat experience. Such differences that existed between the opposing forces in regard to artillery, small arms, and other weaponry were collectively not enough to confer a decisive advantage on either side of the kind the Germans enjoyed over Polish forces in 1939. The equipment of the combatants, then, reflects neither German invulnerability, nor French weakness; one could not predict with certainty the outcome from a survey of the opponents' weaponry.

Nor could one speculate with confidence on the outcome based on the origins of the German war plan, which was corrupted by animosity between Hitler and his generals, as well as by competing agendas among senior military commanders. The original "Case Yellow" that the General Staff submitted to Hitler in late 1939 amounted to nothing more than an uninspired recycling of the Schlieffen Plan, embodying the letter of the plan that Germany went to war with in 1914, even as it was bereft of its ambitious spirit. The Schlieffen Plan called for a massive wheeling maneuver that would envelope Paris and pin the bulk of the French forces against the German-Swiss frontier. The Kaiser's armies would achieve victory by exploiting their own mobility and French military strategy, which was based on an attack through Alsace-Lorraine, far away from the main German effort.

The original draft of Case Yellow was also built around an attack through Belgium and Holland, but its objectives were fainthearted by comparison. Unlike the Schlieffen Plan, Case Yellow did not seek decisive victory; its objectives were to batter Allied forces, create a protective buffer for the Ruhr industrial region, and occupy strategically advantageous territory so that the war could be more efficiently prosecuted against France and England. For all of its timidity, the original draft of Case Yellow carried risks that were not in play in 1914.

For starters, the strategic surprise of 1914 could not be counted on in 1940. The French expected the Germans to come through Belgium and Holland, which was hardly surprising given the heavily garrisoned Maginot Line, the difficulty of traversing the Ardennes Forest, and the precedent of World War I; unlike Schlieffen, Case Yellow would be a frontal, rather than a flank attack. There was also the possibility that an aggressive French commander might marshal forces on the exposed southern flank of the advance and cut its lines of communication just at the moment when the German offensive, worn down by breaking through Allied defenses, was running out of steam.[34]

A more insidious risk lay in trampling upon the neutrality of both Belgium and Holland in pursuit of a military objective of limited value. The Schlieffen Plan weighed the strategic risk of decisively defeating the French Army against violating a treaty that the signatories (Prussia, Austria, France, Great Britain, and Russia) had honored since 1839. By contrast, Case Yellow in its original form would have left the Allies undefeated and might well have provoked the entry of the United States into the war, either as a combatant, or as a supplier of arms and matériel to the Allies.

So large were the flaws in the original Case Yellow that the submission of the plan to Hitler can be interpreted as a form of insubordination, insofar as General von Brauchitsch, commander in chief of the army, and his chief of staff, General Halder, saw nothing but catastrophe in an all-out assault on France before 1942. The point here is that even a well-trained and intellectually gifted staff can produce an insipid war plan—one that was bound to repeat the stalemate that led to Germany's defeat in World War I.

After much debate and bureaucratic maneuvering, envenomed at times by Hitler's contempt for the General Staff (which was reciprocated) and by reflexive misgivings among a few senior commanders about any audacious stroke against France, "Sickle Cut," as the plan came to be known, was settled upon.

The strategic objective of Sickle Cut was not to conquer territory or seize towns, but to destroy the enemy armies in the field, as Hitler's War Directive 10, issued on 20 February 1940, makes clear:

> The objective of offensive "Yellow" is to deny Holland and Belgium to the English by swiftly occupying them; to defeat, by an attack through the Belgian and Luxembourg territory, the largest possible forces of the Anglo-French army; and thereby to pave the way for the destruction of the military strength of the enemy.[35]

Sickle Cut called for the employment of three army groups. The southern army group would face the Maginot Line, thus absorbing the attention of the 400,000 French troops posted there. This group contained no panzer divisions and was the least formidable of the three. The northern army group, which included the weakest three of the *Wehrmacht's* ten panzer divisions, would attack France by way of Belgium and Holland, the objective being to divert the heart of the Allied armies away from the main German blow which was to be delivered south of the Liege-Namur axis. The *Luftwaffe* would concentrate its efforts in the north as a means of disguising further the location of the *Wehrmacht's* main effort. The most powerful of the three army groups, assembled under cover of the Ardennes forest, centered around seven panzer divisions, and was tasked to seize bridgeheads across the Meuse between Dinant and Sedan, and from there to drive for the coast, thus trapping the Allied armies in northeast France and Flanders, thereby separating them from French forces on and south of the Somme.

Intellectually marvelous, Sickle Cut reconciled boldness with prudence. It is a commonplace among staff officers and commanders, as Helmuth von Moltke observed in his history of the Franco-Prussian War, that no plan survives first contact with the enemy—the implication being that once a plan is set in motion, victory depends on nerve and clearheaded improvisation. Sickle Cut took full account of friction, fog, and chance. If all went well for the Germans (as it eventually did), Allied forces, assuming that their southern flank was protected by the impassibility of the Ardennes and the impregnability of the Maginot Line, would move northeast to repulse what to them seemed like the most effective avenue of approach. The army group advancing west across the Meuse River would ensnare these forces in one huge pocket. But if the Allies decided to establish a firm line before counterattacking—a not implausible assumption given their defensive-mindedness—then the fast-moving German armored formations would paralyze Allied command. This would likely happen even if French resistance along the Meuse was intelligently directed, and even if traffic snarls impeded armored columns making their way through the Ardennes, if only because senior French commanders would likely be unable to determine the main line of attack before it was too late. Hardly less important is that the plan accommodated both Hitler's Napoleonic self-confidence and the General Staff's fear of repeating the catastrophe of World War I, which almost all senior commanders had experienced firsthand.

Sickle Cut perfectly balanced strategic objectives against strategic risks. It considered all reasonable possibilities in regard to the enemy's reaction to attack, and forces were composed and allocated in such a way as to match German strengths against allied weaknesses. In fact, so renowned is the plan among military historians that its architect, Erich von Manstein, is better known nowadays for Sickle Cut than for his illustrious achievements as an operational commander on the Eastern Front.

The development of the plan demonstrates the productive interplay between conventional thinking and innovation among the German General Staff. The emergence of von Manstein's ideas was the byproduct of a professional culture that not only tolerated, but encouraged rigorous debate right up until an order was executed. Manstein's plan, no matter how brilliant, would never have seen the light of day had they not been given a sympathetic hearing not only by Hitler (bold ideas were very much to his liking), but also by Manstein's rather conservative-minded commander, General von Rundstedt. Hardly less relevant is that Manstein, Rundstedt, and several other senior commanders embodied the high traditions of the German General Staff. Officers were chosen for duty based on their intellectual ability rather than on their political views, or because advancement in rank absolutely required it. There was no corporate method or formula, nor a bandwagon culture, that might have typecast Manstein's thinking as hopelessly exotic.

The French began planning to repulse a German invasion in late September 1939, at about the same time that Hitler issued War Directive 6: the tasking for the original Case Yellow.[36] By the end of 1939 the French High Command had settled on Plan D (which signified the area near the River Dyle, where French forces, supported by British detachments, would marshal). Because the French assumed that the German forces would most likely drive across the Belgian plain, Plan D concentrated Allied forces and most of their tanks and motorized transport in Northeast France, on a line west of the Antwerp-Namur axis. In the south, the Maginot Line was amply provided with infantry and artillery. Thus, the Allied front comprised two strong wings. Between them was a center that was held by forces deficient in ability, numbers, and equipment. These weaknesses, it was believed, were adequately compensated for by the rugged upland country of the Ardennes, and by the fact that German forces would have to cross the Meuse between Dinant and Sedan—a much more difficult undertaking than establishing bridgeheads on the narrower and shallower rivers in the north.

In devising Plan D the French worked from the following three assumptions. First, the Germans would attack through Belgium and nowhere else in strength. Second, the Germans must not be allowed to occupy French territory; the battle must be won on Belgian or Dutch soil. Third, despite their successful employment of vast pincer movements in Poland, geography would force the Germans to rely on a frontal assault.

It is worth considering the validity of each of these assumptions. The French understood their center of gravity as residing in the country's industrial heartland and in its capital, which could, on account of the flat terrain, good roads, and relatively short distance, be most easily occupied by driving across the Belgian frontier. Advancing through the Ardennes would be foolish, not only because of the uncongenial countryside, but also because the Germans, who, in attempting to drive a wedge between two strong wings, would leave their flanks exposed to counterattack. Attempting to breach the Maginot Line would result in a replay of Verdun, not for the French but for the Germans; this time there would be no Fort Douaumount.

The second and third assumptions, that France must hold the Germans back from French soil even as they decided to wait for the *Wehrmacht* to strike, are partially rooted in the experiences of World War I, when France's misbegotten offensive strategy yielded carnage beyond belief and nearly ended in defeat. But the defensive-mindedness embodied in Plan D also reflected the defeatism that consumed France between the world wars.

In 1940 France sought to avoid a military test of strength with Germany. She had turned a blind eye toward Hitler's flouting of the Versailles Treaty during the mid1930s, and accepted war with Germany in 1939 with conspicuous reluctance. As noted earlier, the unprecedented brutality of the Great War spawned in France a school of pathologies that would paralyze her ability and will to fight in 1940, among which are the following: a stubborn popular indifference to strategic matters, particularly in regard to military funding and conscription, and correspondingly, an ethos of individual pleasure-seeking; the frequent detonation of sharp and sometimes violent political fractiousness, pent up during World War I as a matter of national survival and pride; a military infected by intellectual complacency; and bureaucratic inertia.[37]

The belief that a German advance could be stopped at the French border, and then rolled back, also illustrates the retrogressive thinking that can afflict any victorious army. Even though the Polish campaign demonstrated the lethality of blitzkrieg, the French rejected outright the possibility of being subdued in a similar way. The French believed themselves to be tougher and more resourceful fighters than the Poles. They also were convinced that the *Wehrmacht's* blitzkrieg doctrine was reckless—effective against a feeble, disorganized opponent, but ineffectual when resisted by an enemy whose courage and resolution stopped the Kaiser's armies at Verdun and on the Marne. There were French officers concerned about the rehabilitated German army (Colonel de Gaulle, for one), but their points of view were peremptorily discounted.

Such was the scope and depth of French self-satisfaction that senior commanders actually looked forward to the German attack: the sooner it came the sooner Germany's perfidious ambition would be thwarted by French valor and the war brought to a swift and happy end.[38] Even when French reconnaissance identified a German build-up between the Rhine and Moselle rivers in the early spring of 1940, it was interpreted as an act of strategic deception. That the Germans might take risks that no French commander would dare countenance was never seriously debated.

It is easy to criticize Plan D given the outcome of the battle, but we should not forget that the plan was, by the standards of conventional thinking on operational matters, a competent piece of work. What has been given remarkably little emphasis in the postmortems on Plan D is the failure of the French to ask searching and disinterested questions about the culture of German military leadership. The French assessed potential German action based on inanimate circumstances: terrain, equipment, doctrine, the proximity of France's industrial centers to potential avenues of approach, the material conditions of the Great War, and so on. They also failed to consider the possibility that Germany had learned a great deal from defeat in 1918, and that the leadership in 1940 was of a wholly different cast from that of the Kaiser and his generals.

Had the French taken stock of Hitler's ambition and mental outlook, which not only found expression in *Mein Kampf*, but also in the bloodless conquests that preceded the Polish campaign, they might have been able to predict with greater accuracy the German course of action. Hitler was a gambler. The French generals might well have asked, given the circumstances, how will a gambler likely behave? What is the best way to thwart a gambler who relies on men of caution (e.g., Generals Brauchitcsh, Halder, Kluge) to achieve his ends? The French commanders seemed largely unaware of the enmity and political rivalries that beset Germany's political and military leadership—weaknesses that, had they been properly understood, might have been exploited once the battle began. Who can know what effects a sharp setback, actual or perceived, on the right bank of the Meuse might have done to the morale of senior German commanders, and, correspondingly, Hitler's resplendent but insecure standing as a military genius?

The character and temperament of the German leadership could not ever have been understood by our system-of-systems approach, embodied in doctrine and exalted in our transformation policy, if only because there was no decision-making process. Hitler's strategic and operational decisions were at times influenced by any number of people; often they were drawn entirely from his own counsel. Decisions made by the generals were hardly less impermeable to formulaic analysis. Who might have foretold that Guderian's hard-driving panzer columns would stop abruptly, just at the moment when a dash up the coast might have annihilated the British Expeditionary Force at Dunkirk? On this point in particular the system-of-systems approach is disquietingly fallible.

The interplay of circumstance and moral factors is illustrated by the British counterattack near Arras on 21 May 1940. The day before, panzer spearheads had reached Abbeyville, which lies just east of the Somme estuary and about thirty-five miles west of Arras, thus cutting France in two. The better part of the Allied forces were trapped in a huge pocket bounded by the sea to the west and north, and German forces to the south and east. Paris and the rest of the country were left virtually undefended; the Maginot garrison was a remote island bereft of the possibility of succor. With German forces established on the coast, the field commanders were preparing to shove a dagger into the southern flank of the allied armies withdrawing to the west, and collaterally, to begin driving north toward Boulogne, Calais, and Dunkirk, which would seal off the English Channel.

The commander of the British Expeditionary Force, John Standish Surtees Prendergast Verecker (Viscount Gort), strenuously worked to coordinate a counterattack with the French, but his efforts came to nothing. Aware of the impending collapse of Belgium, and confounded by the rapidity of the German advance, the French High Command had already come to terms with an inevitable defeat. Their conduct brings to mind a large and powerful boxer, who, having received a concussive punch to the head in the opening seconds of round one, can muster enough strength merely to stagger to the center of the ring in the hope that the referee's whistle will forestall a lethal blow. It appeared that the French had despaired of the possibility of victory or even stalemate, but carried on in the hope that their country would be spared the fate of Rotterdam and Warsaw.

On 21 May the British attacked, their objective being to sever the German lines of communication before the infantry could move up and seal the corridor to the Atlantic. The British took heavy casualties (a few at the hands of French antitank gunners, who mistook the Matilda tanks for German models) and the advance ground to a halt after covering about ten miles, from just west of Arras, to about five miles to the southeast. Along the way, one of Hitler's crack divisions, *SS Totenkopf*, absorbed heavy casualties and was put to flight. For once, the soldiers who surrendered in the hundreds were dressed in *feldgrau*, and the burning tanks and trucks carried the *Balkenkreuz* on their sides. Rommel's panzers could not stop the British Matildas, nor could the shells of the 3.7 cm PAK antitank guns. Only Rommel's morale-restoring intervention at the front, along with the tank-destroying power of the Krupp-made 8.8cm flak guns, managed to save the day for the Germans.

Hopeless though the British counterattack may seem by the lights of history, the impact on the mind of the enemy at the time was astonishing. Rommel, whose cool-headedness in battle is on display in his World War I memoir, *Infantry Attacks!,* grossly overestimated the British force. His map of the battle shows the British attacking with five armored divisions. German forces in the field remained unshaken; most of the army, even at that time, had seen little actual fighting. But the brain of the army, if only momentarily, panicked. Guderian's hard-charging panzer corps was ordered to put one of its divisions, the Tenth Panzer, in reserve. The German Fourth Army refrained from advancing further until the situation was restored. The Sixth and Eighth panzer divisions ceased offensive operations and took up defensive positions along the flank of what was assumed to be a concentration of allied armored divisions. In a word, the Germans had been stopped by their own fears, which helped to fabricate a respect for the enemy that could not be justified by fact and recent experience. As for the British, the effect of the attack was, if anything, demoralizing. Viscount Gort gave up hope that the Germans could be stopped, and began arranging for the retreat and eventual evacuation of the British Expeditionary Force.[39]

The war between France and Germany puts the lie to the idea that an enemy is best understood as a system, that the prosecution of war can be made to conform to the template of a business enterprise amenable to such catchphrases such as, "right weapon, right place, right time," and that a strategy blind to all but abstractions can generate antiseptically predictable results. Hard strategic considerations, and a thorough, disinterested understanding of the cultural temperament of the belligerents must always take top priority—something that becomes apparent when one considers the interplay of combat operations and the strategy it was intended to realize.

The brilliance of the design and execution of the campaign aside, Germany's strategy was rotten at the core. No British self-interest could have been served by coming to terms with a resurgent Germany, armed to the teeth, and established on the far side of the British Channel. So long as Germany dominated the whole of Europe the United Kingdom would remain her antagonist. Given her imperial possessions and her close ties to the United States (culturally, if not always diplomatically), the best that might have been accomplished would have been to neutralize Great Britain's ability to obstruct German strategic aims by controlling the Baltic, eliminating Russia, and exploiting the long-standing, if often latent, estrangement of the French and British. The German armed forces were best suited for fighting expeditionary campaigns against continental opponents; the naval and air components were adequate, just barely, to support such a strategy. Waging war successfully against the United Kingdom would have required a maritime strategy built around the ability to conduct littoral operations, something that was alien to the German experience. To defeat Great Britain also required the wherewithal to fight on at least two fronts: the Middle East, as well as the region bounded by the Baltic, the North Sea, and the Atlantic. And such a campaign would have had to take intelligent account of the strategic sinew, bone, and heart of the United States.

For the Germans, the occupation of France ultimately became a strategic liability that required the deployment of troops the *Wehrmacht* could not spare once war with Russia commenced. The Battle of Britain was a failure because that campaign was based on the notion—one that survives today—that a defiant though militarily enfeebled enemy can be brought to terms by the employment of airpower alone, an early example of effects-based operations. Had Germany crafted a magnanimous peace with Poland and then destroyed the Soviet Union—along the way liberating, rather than enslaving, the Baltic states and the regions of the western Soviet Union—substantial portions of the population in the west would have reacted not unsympathetically, especially the French, who had long wrestled with indigenous communist troublemaking.

But of course, all of this is impotent speculation, because Hitler was motivated not by rational strategic calculation, but by the desire for revenge (against the French in particular), and, much more powerfully, by a racial ideology that was as stupid as it was an embodiment of an iniquity that was rivaled only by the ambitions and practices of Josef Stalin. In recognizing Hitler as an avatar of evil, the Western Allies were belatedly and only partially right; they were themselves not entirely clearheaded in their own strategy. Understandably, the British saw the war after the French capitulation as a struggle to remain free, a view that Hitler basically forced upon them despite his peace overtures. From December 1941 onward the United States similarly viewed the war as pitting democracy against Hitlerism, with Stalinist Russia as a stalwart and much-admired ally.

Even today, to the extent that the Second World War is remembered in the United States, it is recalled, without qualification, as the "good war"—a characterization that is true only in a limited sense. The simpleminded outlook of the western allies shaped their strategy, led ultimately to Soviet domination of Eastern Europe, and brought in its wake the fifty-year-long Cold War. For the Allies, it would have been a straightforward matter to pursue the war not merely with the eradication of Nazism as its object, but also with the aim of preserving European civilization from the awful oppression of Stalin and his successors. Berlin, Prague, Vienna, Budapest, and perhaps even Warsaw might have been saved by a strategy conceived with an unromantic understanding of Soviet communism. As J. F. C. Fuller has argued, had the Allies taken diplomatic steps to express support for the latent anti-Nazi feelings among senior German officers, the war might easily have ended in 1943 with Hitler dispatched to meet his reward in eternity, and his moral peer, Stalin, contained.[40]

Thus, the war between France and Germany illustrates for us the inviolable bond that unites operational and strategic thought. When planning for conflict, we must always begin with a clear idea of the peace we are seeking, and then build the forces and write the plan that is most likely to realize that end. The Germans recognized in France the principal impediment to European domination, and Hitler's appetite for revenge made war between the two states inevitable as well. But military supremacy can yield self-wounding habits of mind. Hitler drew the wrong lessons from the French campaign, mistaking victory over a morally feeble opponent, for invincibility. On the other side of the hill, France crouched behind her Maginot Line and took empty solace in her past cultural and military achievements. She, too, succumbed to vainglory, which for her was a means of medicating a military impotence that any clear-eyed strategist could see was heading for retribution. As we ponder our own position in the world, we ought not to ignore the strategic outlook of two military powers that collaborated in their own destruction.

Chapter 5

Stalingrad

For all of its squalor and brutality, the battle for Stalingrad does not stand in relief when compared with other sanguinary struggles of the twentieth century, or even with other battles fought between the Germans and the Russians during the Second World War. Operation *Bagration*, the Soviet offensive that coincided with the invasion of Normandy, was far more costly to the Germans. Between July and September 1944, the various Soviet offensives imposed on the German army nearly a million casualties; more than one hundred divisions were effectively removed from the *Wehrmacht's* order of battle. Tens of thousands of German prisoners of war—gaunt, broken men dressed in rags—were paraded through the streets of Moscow in July. The subjugation in 1945 of the fortress cities of Königsberg, Breslau, Danzig, Gotenhafen, and Berlin, and the sacking of East Prussia rival in mercilessness and fury, if not in strategic importance, the battle for Stalingrad. The city, like most of those previously mentioned, no longer bears the name it did sixty years ago. Volgograd began appearing on maps of the Soviet Union in 1961—an expression of Nikita Krushchev's de-Stalinization initiative that also provided the occasion for the publication of Georgy Zhukov's memoirs, the most reliable (if not entirely truthful) of those written by Soviet officers in the aftermath of the war.[1]

Yet even today Stalingrad remains a particularly evocative word. When the United States invaded Iraq in 2003, news writers and television pundits could hardly restrain themselves in predicting (the warnings soon took the form of a litany) that the taking of Baghdad might well turn into another "Stalingrad," an example among a multitude of examples of the antihistorical outlook that has permeated the modern mind. Online auction sites have transmogrified Stalingrad into a wellspring of kitsch. One can buy a ballpoint pen, complete with key chain, made out of a German machine-gun bullet found in Volgograd for ten dollars. A recently excavated aluminum spoon-and-fork tool commonly used by German soldiers goes for around forty dollars. The beginning auction price for a Romanian officer's teapot is one hundred dollars, even though prospective customers from Germany, Italy, and France are not allowed to bid. Mass-market recording companies peddle compact discs that feature eponymous songs (there's even a *Stalingrad Symphony*), presumably as a means of exploiting the appetite among adolescent boys for anthems celebrating nihilistic destruction conveyed by truculently giftless musicians. In 1993 a movie was made about the battle, centering on the experiences of a platoon of German combat engineers. In many ways an outstanding film, *Stalingrad* reflected the self-reproach among modern Germans in regard to the Second World War, more than it offered an evenhanded portrayal of the battle. For that, one must read William Craig's *Enemy at the Gates* and Antony Beevor's *Stalingrad: The Fateful Siege,* two engagingly written books that will remain standard works on the subject.

In many ways Stalingrad was a distinctively modern campaign, though the scale of battle is scarcely imaginable to devotees of network-centric warfare, who believe that war is best understood as a contest between computers served by technicians in uniform. Directed from afar by their respective heads of state, hundreds of thousands of well-equipped troops, along with many hundreds of tanks and scores of bombers and fighter planes, clashed over a period of six months in the service of that conspicuously modern weapon, ideology.

Traumatized though they were by the fighting during the previous winter, the German army in the spring of 1942 was in some ways a more effective force than ever before. The *Wehrmacht's* newly acquired skill at conducting mobile defense, combined with its customary verve on the offensive, lately fortified by up-to-date weaponry, made it the most powerful army in the world. Between August 1942 and February 1943, however, the *Wehrmacht* absorbed a battering from which it never recovered, not because it was outfought, but because the German head of state, apathetically supported by the General Staff, placed the *Ostheer* in one untenable position after another. When the taking of Stalingrad became a matter of prestige, it was no longer of any consequence to Hitler that his panzer divisions and their supporting infantry were employed in ways that marginalized their strengths, and played to the strengths of the enemy. Hitler imbued Stalingrad with an ideological significance that could never be reconciled with commonsensical strategic priorities, nor with the unyielding geographical and material circumstances of the war.

At some level Hitler seems to have accommodated himself to the destruction of the Sixth Army long before the shooting ceased in and around Stalingrad. For him, a fitting consolation was that the annihilation of the Sixth Army provided ample material for National Socialist mythmaking, a point validated by the generous bestowal of medals and promotions on the encircled army days before its collapse. Paulus surrendered on the day Hitler promoted him to field marshal, on the assumption that Paulus would shoot himself rather than go into captivity. In discussing the defeat afterward, Hitler seems to have been far angrier with Paulus for refusing to commit suicide, than remorseful, or even discomfited, by the huge losses in men, equipment, and strategic position. In this regard, the battle for Stalingrad illustrates as well as any other that advancements in technology and tactics do not necessarily lessen the primitive core of war, nor do they eclipse war's essence, which is at heart a test of moral and intellectual strength.[2]

Put another way, war is heavily dependent upon confidence and psychology. As good as the *Wehrmacht* was in the summer of 1942, its most effective weapon was the deeply ingrained sense of inferiority its successes had inflicted on the Russians. Neither Hitler nor his commanders exploited this potent force. From the start of the Stalingrad counteroffensive, until the Sixth Army was isolated beyond hope of rescue, the Russians were vexed by the possibility of a German counterstroke. Indeed, that the Russians massed seven armies around the pocket is an expression of their underlying fear of German operational excellence. Had the Germans employed a mobile reserve against the pincers moving to encircle the Sixth Army, they might well have transformed what would have been a modest reversal for the Red Army into a rout.

This nearly happened on the afternoon of 21 November 1942, two days after the Russian counteroffensive began. The Twenty-ninth Motorized Infantry Division, which had built a distinguished combat record over the previous eighteen months, had been pulled from the line and was resting about ten miles south of Stalingrad, when it was ordered to attack Red Army forces that had broken through the Fourth Romanian Army guarding the southeastern flank of the Sixth Army. On its move southward, the Twenty-ninth smashed into the right flank of the Soviet Fifty-seventh Army. The division's tank battalion ripped to shreds the Thirteenth Mechanized Corps, which began the fight with some ninety T-34s. The panzer grenadiers, mounted in armored semitracked vehicles, shot up the disorganized Russian infantry and their supply columns. At about the same moment, a Russian armored train appeared, followed by other trains carrying infantry. These became targets of opportunity for the Mark IV tank crews, who began firing "hundreds of shells into the packed boxcars. Through binoculars, the gunners watched countless Russian bodies cartwheeling into the air and down onto the snow"[3] General Leyser was in the midst of organizing his division to deliver an annihilating blow, when, upon receiving direction from Army Group B to turn northwestward to protect the Don River crossings, he was forced to abandon the attack. But even this limited success knocked the Russian leadership off balance. General Viktor Volsky, commander of the Soviet Fourth Tank Corps, which provided the armored punch for the Fifty-seventh Army, slowed down his drive in order to await reinforcements.

General Leyser's commanders, 200 miles away, thus prevented him from exploiting a brilliantly conducted battle of encounter between the Twenty-ninth Motorized Division and the Soviet Fifty-seventh Army, proving that fog and friction are as often as not self-inflicted. Here we have an example of a small force, led with considerable dash, that imposed a demoralizing setback on a much larger foe. Had higher headquarters, which is supposed to have the broad picture, apprehended Red Army intentions, the Twenty-ninth might have gone on to induce panic among senior Soviet commanders who no doubt remembered the dark days of July 1941. It is more than a footnote to the Stalingrad campaign to recall that on 12 January 1943, General Leyser's division, hunkered down on the ice-blasted steppe, managed to repel ten to twelve Red Army divisions and destroy some hundred tanks in a single day, before being liquidated. As the Russians began their last offensive to destroy the Sixth Army, on the western edge of the pocket only the decimated but unshaken Twenty-ninth Motorized Division stood in their way.

> Gen. Ernst Leyser was there with his men, urging them from the houses and holes in which they cowered. He had only four tanks left but after dark, he broke cover and ran forward screaming: 'Hurrah!' Hundreds of wounded and previously apathetic Germans suddenly jumped out to follow Leyser in an attack against the surprised Russians. The firefight was a confusion of men and shellbursts, dead, and more wounded crying in the blood-stained snow. Leyser had thrown the enemy off balance for a moment. . . .[4]

It's worth recalling also that this last stand took place more than a month after the Sixth Army recorded its first starvation case. The stalwart performance of the remnants of the Twenty-ninth Motorized Division cannot be explained by the courage of despair, or by an ardent faith in National Socialism, or even by confidence in the *Wehrmacht's* senior leadership. Rather, it was the fruit of an inspirational commander who managed to eke out the last measure of élan from troops, who under the circumstances might have been expected either to surrender or flee for their lives.

Putting aside for a moment the spectacular bravery of commanders and their soldiers, the human cost of the Stalingrad battle was atrocious, and the aftermath outstandingly gruesome. On Hitler's orders, the fight against Russia was to be a war free from chivalric code or impulse, something that did not particularly trouble the Russian troops from the start, if the mutilated corpses of German soldiers taken prisoner in the opening days of the war is any indication.[5] At Stalingrad, the unyielding viciousness that Hitler demanded, and that Stalin by nature inclined toward anyway, reached glorious perfection. Much of the fighting within the city was done hand to hand, or between the sniper and his unwitting victims. Quarter was neither expected nor given. Reading firsthand accounts of the fighting in Beevor and Craig brings to mind the combat spectacles in *The Iliad*, bereft of even a trace of nobility and honor. There is something memorably, sickeningly pathetic in the scenes at the German-held airfields in the last weeks of the battle. Generally, a life-threatening wound was not something to be feared but hoped for, as it represented the best chance to be evacuated. At other times, a serious wound might justify removing a soldier from an evacuation flight, because stretchers took up the room of three or even four lightly wounded soldiers, who would later help fill the *Wehrmacht's* ranks that had been hollowed out by the impending destruction of the Sixth Army. It became customary to assume that soldiers with hand or leg wounds had shot themselves (often powder burns proved that they did), and these "enemies of National Socialism" were executed on the spot by military police. Knowing this, a few desperate men put a bullet in their stomach, avoiding powder burns by firing through a loaf of bread or a handful of rags. Sometimes the wounds were mortal, but in a few instances soldiers successfully got away with it and received a hero's welcome at home.

The airfields provided scenes of a military version of Noah's ark. Specialists were flown out of the pocket in order to form cadres around which the formations destroyed in Stalingrad could be rebuilt—only to be crushed later on by Russian forces that well knew how to exploit Hitler's strategic follies. As early as the first week in December 1942, German soldiers began dying of malnutrition. This trend was reported on autopsy reports as "death by disease" or the like, thus demonstrating that the battle managed to corrupt even those who were sworn to uphold the Hippocratic oath.[6]

The civilian population of Stalingrad was not evacuated; Stalin believed that the soldiers would fight more ferociously for a populated city. When the shooting stopped, only about 1,500 civilians (living amid, but usually beneath the rubble) out of a population of about 500,000 survived. Russian military casualty figures are not particularly reliable, but an unofficial estimate provided to William Craig is three-quarters of a million dead, wounded, and missing—a number that is probably low. The Hungarian, Italian, and Romanian armies were destroyed. The German army was crippled, not only by the losses of equipment and territory, but also in terms of military and civilian morale. The largest and best formations of the *Wehrmacht* were annihilated at Stalingrad. More than a score of German generals were killed or captured. For loved ones at home, the Sixth Army disappeared without a word as to the fate of individual soldiers. In the spring of 1943 a handful of letters from survivors reached Germany through Turkey. When asked for permission to deliver them to families, Hitler, with an astonished expression draped across his face, refused. "The duty of the men of Stalingrad is to be dead."[7]

About 500,000 German, Italian, Hungarian, Croatian, and Romanian soldiers marched into captivity. Roughly half of the 91,000 Germans who surrendered when the shooting stopped, perished within a few weeks of malnutrition, typhus, and illnesses relating to combat wounds. Many others were murdered by Russian soldiers who interpreted falling out from the march columns as a death wish. Those who survived ended up in a variety of slave labor camps. By April 1943 more than 400,000 Axis prisoners had died. Beginning in May 1943, the Russians began to feed their prisoners at something approaching a subsistence level, but some of the survivors still turned to cannibalism. Only about 6,000 German prisoners ever returned home. The last ones were repatriated in 1955, thanks to diplomatic pressure applied by Konrad Adenauer, chancellor of the West German government.[8]

Occupying Stalingrad was at first not a principal objective for the Germans. But Hitler soon became hypnotized by the symbolism of conquering a place that bore the name of his nemesis, and perhaps, also, by the sheer ferocity of the conflict. Titanic struggles were a glorious thing for Hitler insofar as they produced National Socialist "heroes"—a frame of mind aggravated by a wooden-headed insistence that ground taken by a German soldier must never be given up. Stalin was hardly less beguiled by the struggle for the city. He had accumulated enough forces to maintain pressure on the central and northern sectors of the Eastern Front, which was an option not available to Hitler, whose forces were absorbing Germany's dwindling industrial resources, as well as her shrinking reserves of manpower.

With its converging rail lines, Moscow in the autumn of 1942 was the most suitable base of operations for conducting a counterattack against the Germans. Stalin might easily have launched a strategic offensive on the central sector along the most direct route to Berlin, which would have shortened his lines of communication and forced the *Wehrmacht's* northern and southern army groups to reverse their frontages. An offensive along the Moscow-Smolensk axis would have threatened the entire Germany army in the east, and might well have paralyzed the German High Command, who would have been hard pressed to determine their ultimate objective before they could begin shunting forces from other sectors. But Stalin seems to have reveled in the idea of delivering a blow that humiliated Hitler, even if it came at the expense of achieving victory in the briefest span of time.[9]

Shortly after the Russians closed the ring around Stalingrad, the German High Command drew the Eleventh and Seventeenth Panzer Divisions from the central sector in an attempt to dam up the hemorrhaging front between the Don and the Volga. The terribly understrength Seventeenth provided flank protection for the newly rehabilitated Sixth Panzer Division as it attempted to relieve the Sixth Army from southwest of the city. About one hundred miles to the north, the Eleventh Panzer Division reached the front just in time to break up Soviet armies that would otherwise have driven, practically unopposed, from the Chir River to Rostov. Balck's panzers and the accompanying motorized infantry formations did no end of damage to Red Army forces, destroying hundreds of tanks in battles in which his forces were always outnumbered, across a battlefield that was, by Eastern Front standards, relatively confined. It is unlikely that the division would have been quite as effective had it been employed against an attack out of Moscow across a much broader front.

In 2003 the impending struggle for Baghdad was not infrequently referred to in press accounts as "Mesopotamian Stalingrad." Given the popularity of this trope, and given the modern cast of the battle, it is tempting to interpret the Stalingrad campaign by the lights of our own current obsessions, that is, in an aridly mechanistic way, using the principles of war and the pseudoscientific terms of operational art, as a means of explaining what can only be understood as the most intense form of social activity, as war always is.

In examining the battle as it might be taught in professional military education programs (particularly at the Joint Forces Staff College), we might note, for example, that in the Stalingrad campaign the Germans went to the offensive from the start, and kept on attacking until their forces were incapable of doing so, a point reached shortly before the Russian counteroffensive began in November 1942. Certainly, the Germans employed "Mass" in Stalingrad: entire army corps were used to attain tactical objectives within the city itself. The entire campaign was nothing, if not an example of economy of force taken to an extreme; a glance at the map and the German order of battle makes this point clearly enough. Less capable allies (Hungarian, Romanian, and

Italian troops) occupied supposedly quiet areas on the flanks of the Sixth Army, thus exemplifying the value of using coalition forces to buttress the war effort. The Germans also employed "Maneuver," using concentric attacks on Stalingrad until the Russians—badly mauled by months of close combat—were left holding only a tiny fraction of the city. The Germans demonstrated "Unity of Command" to a fault, as Hitler denied his field commanders independence of action. "Security": in early November, Sixth Army intelligence identified the Russian troop build-up on the west bank of the Don, north of the city, though commanders at the strategic and operational levels badly underestimated Soviet intentions and the robustness of their strategic reserve. Here we might say that this exemplifies the necessity of better intelligence gathering resources. As for "Surprise," the entire German operation caught the Soviet High Command (the *Stavka*) off guard. In the summer of 1942 Stalin was expecting an attack on the central front, with Moscow as its objective. This assumption was in some ways endorsed when a German reconnaissance plane, carrying staff officers with the operational plans in briefcases, crashed behind Soviet lines, thus unintentionally illustrating the value of strategic deception—part of what we call nowadays "information operations."

Hitler's war directives that bear on the Stalingrad campaign are a marvel of lucidity and concision: "Simplicity" exemplified. Swayed by his economic advisors, Hitler believed that the war could be won by capturing the oil and grain of the Caucasus, and the region bounded by Voronezh on the north, Rostov on the south, and the Volga River to the east. Field commanders who noted that the objectives were too dispersed to be attained with the troops available for the operation were intimidated into silence. The unfavorable ratio of force to space was not lost on the General Staff, which recommended an additional draft of 800,000 troops to fill out the ranks for the spring offensive; but Albert Speer, Hitler's economics minister, insisted that the war industries could not spare them. More compliant generals, such as Friedrich von Paulus, were promoted and given wider authority. Hitler's optimism, moreover, was fortified by the delivery of new weapons leaving the factories—the PAK 40 7.5 cm antitank gun, and the upgunned Mark III and Mark IV tanks—to frontline troops. Technology, if not numbers, seemed to have moved in favor of the *Wehrmacht*, which despite the heavy losses of the previous year was still tactically superior to the Red Army, though the gap at the operational level of command was rapidly closing.[10]

What went wrong for the Germans? The Soviets exploited fully their reserves of manpower and equipment, including large numbers of T-34 tanks, thereby employing greater mass at decisive points to realize their objective of entombing the Sixth Army. "Surprise" was achieved by the double envelopment. The Germans did not expect the attack on their southern flank. "Economy of Force" was realized when the Russians supplied the armies fighting in Stalingrad with just enough reinforcements to keep the city from being overrun—a forerunner of our own "just-in-time logistics" practices, and an early example of "off-ramping," that is, avoiding an excessive build-up of forces. The *Stavka*'s plan was simple enough and Stalin, unlike his counterpart, did not interfere unduly in operational matters. We might look at this as a validation of horizontal decision-making structures. A senior officer and staff from General Headquarters in Moscow superintended the three front commanders (a front was basically an army group) involved in the Stalingrad operation, thus eliminating intermediate levels of control.[11]

Because both sides employed the principles of war and the tools of operational art in imperfect but nominally effective ways, what lessons about Stalingrad is the fourth generation warrior likely to take from such a drill? The temptation, perhaps the natural course of this kind of exercise, is to place these items on a scale which will tip toward the Russians, because their use of mass, maneuver, security, and surprise was ultimately more decisive than that of the Germans. By contrast, the Germans worked from an imperfect understanding of effects based operations. Bombing Stalingrad merely turned buildings into rubble fortresses and strengthened, rather than undermined, the will of the Soviet combatants. Soviet effects-based operations were more successful. Early on the Red Army learned the psychological impact of the sniper. In the counterstroke that entombed the Sixth Army, the Russians attacked the ill-equipped and poorly motivated Romanians on the flanks, thus achieving their ends in remarkably good time and with a minimum of casualties. In fact, the pincer attacks could be said to illustrate asymmetric warfare. Instead of attacking the Germans in the immediate vicinity of Stalingrad, the Russians exploited their strengths (mobility and mass) against the weak allies of the Germans, whose infantry units were unsupported by tanks, artillery, and motorized transport.

Such an interpretation of the campaign, though based on fact, is of little use. Of course, the battle for Stalingrad was much more complicated and thus offers us an opportunity to reflect on the timeless essence of strategy.

Military strategy must begin with a disinterested and painstakingly reasoned reconciliation of ways, ends, and means. Collaterally, a great deal of thought must be devoted to a consideration of the enemy's moral and physical resources, as well as the kind of peace he is ultimately seeking. When measured by these standards, Hitler's Stalingrad plan was self-injurious from the start. Sensibly enough, given the collapse of Operation *Barbarossa*, Hitler in early 1942 became increasingly concerned with the raw materials of war, especially oil. Seizing the oil fields in the Caucasus seemed the only way to solve this vexing problem, or so he thought. Of course, this is 180-proof folly. Strategic materials, including a generous reserve, must be secured before starting a war. Napoleon's armies managed by pillaging the countryside as they went, but mechanized forces can't be sustained in such a way. (One of the minor, but nonetheless menacing, frustrations for the panzer columns advancing across Russia in 1941 was the absence of gasoline stations that had been conveniently available in French towns the year before.) Worse yet, the huge losses in men and equipment in the aftermath of *Barbarossa*—several panzer divisions were bereft of their entire tank complement, and almost all the infantry divisions would never again attain full strength—meant that Hitler had to rely on less capable allies to fill out his legions. He also had no choice but to concentrate his forces in the south, without the urgent and necessary ability to apply pressure on other sectors. The campaign would thus become a contest between Germany's shrinking pool of manpower and industrial capacity—not to mention the permanent, if temporarily recessed damage to morale which the Soviet winter offensives had effected—against the accumulating moral and material strength of her adversary.

There were other signs of gathering catastrophe before Hitler unleashed his panzers in June 1942. "Case Blue," the codename for the summer offensive, called for the employment of two army groups. The southern force, Army Group A, would seize Rostov on the Black Sea in coordination with a separate offensive by Erich von Manstein's Eleventh Army, tasked to expel the Red Army from the Crimea and the Kerch Peninsula.[12] With Rostov as its base of operations, the armored spearheads of Army Group A would drive south, clearing the Kuban peninsula, enroute to capturing Baku on the Caspian Sea and Batumi on the Turkish border. Army Group B, meanwhile, would push out of from its assembly area north of Kharkov and strike southeastward toward the right bank of the Volga, securing the northern flank of Army Group A, and cutting off industrial traffic from the south by bringing Stalingrad and northbound river traffic under artillery fire. The tentative sequel plan called for an offensive out of the Stalingrad area in the direction of Kazan, where the Volga bends from east to south, followed by an assault on Moscow, and the isolation of Kuybyshev and the newly constructed industrial centers west of the Ural River. The strategic weaknesses of the plan of attack were of two kinds.

First, Case Blue created insanely extended flanks in support of forces that were absurdly small, given the territorial objectives (incredibly, Manstein's Eleventh Army was transferred to Leningrad in July 1942), a risk we seem willing to take today, insofar as we seem bent on reshaping our military along the lines of a business firm devoted to lean inventories. No army executing an ambitious offensive can be strong everywhere, least of all the much-diminished German army of 1942. It would be a simple matter for the Russians to allow the *Wehrmacht* to expend energy in the drive eastward, granting them the psychic compensation of pursuing an enemy who has chosen to vacate the battlefield, and then probe for a suitable area for strategic counterattacks. This is exactly what the Russians did throughout the autumn on the southern sector. They also kept up pressure with a series of counteroffensives—some of them costly failures—on the central and northern sectors of the Eastern Front. The fate of the entire southern wing of the German armies in the east, not to mention the outcome of the war, was sealed once the *Stavka* became convinced that Moscow faced no threat, and that the Germans had gambled away their vigor in attempting to capture Stalingrad and the Caucasian oil fields.[13]

The underlying and much greater blunder was the assumption that total victory could still be achieved in 1942, and that it hinged entirely on acquiring a single strategic resource. Even if the oilfields had been captured intact (the retreating Russian forces destroyed most of them) there remained the problem of moving the crude product to refineries hundreds of miles to the west. Partisans—well-armed irregular forces operating with impunity from the huge areas left unoccupied by the advancing *Wehrmacht,* and motivated by hatred of the atrocity-committing invader, and in some cases, by loyalty to Stalin and love of country—imposed increasing burdens on the lines of communication. Germany was also wholly unprepared to fight a war of attrition, but that was exactly the sort of conflict she courted once Hitler acknowledged that huge risks had to be accepted to acquire oil. What is more, by 1942 the Soviet armed forces were receiving large amounts of material from the United States and Great Britain. Vehicles and weaponry comprised the largest share of imports, but other, hardly less vital supplies poured into the Soviet Union via Persia and Archangel. Welding equipment, which allowed for the repair of damaged tanks, radio sets and communications cables, canned food, and clothing also reached the Red Army in great abundance. These things, along with commanders who learned a great deal from defeat at the hands of the Germans, conferred on the Red Army a strategic mobility that the Germans never enjoyed on the Eastern Front.

The upshot was that Operation *Barbarossa* led to the reform of the Red Army that otherwise might never have been accomplished. Stalin had the best of two worlds by the summer of 1942: a large, well-equipped army, led by a new generation of talented commanders such as Marshal Rokossovsky, who knew well the penalties for not meeting

Stalin's expectations. Stalin also enjoyed absolute dominion over his subjects, achieved partly by German atrocities, and partly by Stalin's own propaganda machine that made sure the conflict was understood not as one of competing ideologies, but as "The Great Patriotic War."

What other options were available to Hitler in the spring of 1942? The appearance of the T-34 in strength, and intelligence reports that Russia's manpower reserves and industrial capacity were much greater than first thought were dismissed as products of the unduly pessimistic imagination of officers not imbued with National Socialist zeal. In November 1941, a few weeks before the Russian counteroffensives in front of Moscow, two of Hitler's three army group commanders—Rundstedt (South) and von Leeb (North)—advised a complete withdrawal to the original starting line. Hitler sacked them; the commander of Army Group Center asked to be relieved on grounds of illness. Their replacements, besotted perhaps by the glitter of promotion, were less willing to voice opposition, and the cowed General Staff dared not encourage the expression of dissenting opinion.

At the beginning of 1942 there was merit in falling back to the June 1941 border, even though a strategic withdrawal carried plenty of risks, and would hardly have resulted in an end that would have gratified the authors of the Versailles Treaty. Germany might have held on to the culturally sympathetic Baltic countries, and built up an impregnable defensive front backed by a formidable mobile reserve, thus shortening their lines of communication and occupying territory more suitable to their limited resources. A strong defensive front might have discouraged Stalin from rejecting an armistice, which was not unlikely given that Stalin always believed that the greater threat to his regime was internal. The ignominy and suspicion that followed Russian soldiers after even brief contact with Western Allies—very much intensified if one had spent even a day as a prisoner of war—illustrates the point. Scarcely less important is that a drive on Berlin at that stage of the war would have lengthened the Red Army's lines of communication against a foe that, in the months since the Russian counteroffensive, had learned how to conduct a mobile defense.

What is more, a stalemate with the Russians might have undermined the Nazi regime. Hitler, weakened by the loss of prestige and the evaporation of the aura of strategic infallibility that he built around himself, might have overreached domestically in order to hold on to power, and ended up being deposed by the various cabals that had tried to eliminate him since the 1930s. A coup d'etat might have spurred diplomatic moves to negotiate with the successors to the Nazis, encouraged perhaps by a militarily aggressive Soviet Union. The legitimate governments of the occupied countries might then have been restored by political rather than military means. To indulge a bit further in uninhibited speculation, with Hitler gone the NATO alliance might have coalesced many years earlier, and opposed a Soviet Union whose border would have ended hundreds of miles to the east of where they were drawn in 1945.

The larger point here is that the desire for national honor, which is often a consuming motivation among heads of state, is always self-defeating when it refuses to be tempered by hard and cautious—up to the point of pessimistic—interpretations of fact. With German troops entangled in Stalingrad and with no reserves to be had, Hitler feared humiliation more than he did catastrophic defeat; in refusing to take seriously the possibility of one, he courted both. The aspiring head of state who makes himself a keen student of Machiavelli and Montaigne (the latter's essay, "Of Glory," comes immediately to mind) has taken steps to avert disaster. Self-governing peoples must always be on the watch for signs that face-saving has eclipsed national self-interest in the shaping and execution of policy. The flip side of this is that the allure of honor divorced from interest is a weakness always worth exploiting, as Stalin's handling of the Stalingrad campaign proves.

It is tempting for us to infer from Hitler's blundering that the opinions of senior military officers should never be discounted, but this isn't quite true. Before the Stalingrad disaster the German people enthusiastically supported Hitler, as did most of the rank and file of the armed forces. The German generals who furtively discussed plans to remove Hitler were entertaining an act that would have been deeply unpopular among citizenry and soldiery alike. And in any case, Hitler's contempt for military opinion had paid big dividends up through 1941. There is, moreover, no evidence that senior commanders, particularly those who received field marshal batons after the defeat of France, considered resigning before the war with Russia commenced, even though Operation *Barbarossa* was deeply flawed. Nor were there any resignations during the summer of 1941 when the atrocious abuse of Russian citizens and prisoners of war by the Nazis should have proved beyond doubt that officers taking orders from Hitler were de facto war criminals. So long as the *Wehrmacht* piled up victories, the generals were content to avert their eyes from the political objectives that were being furthered. The verdict of history, then, is that the achievements of individual German generals must always remain disfigured by the regime they chose to serve—a decision that cannot be reversed or mitigated by rebellion that coincided with strategic failure.

The moral intelligence of senior *Wehrmacht* commanders often compared unfavorably to that of the soldiers they eventually betrayed, though there were conspicuously noble exceptions. The conduct of General Ewald von Kleist, who took command of Army

Group A after Stalingrad was surrounded, was exemplary. Kleist's fame justly stands on extracting some twenty divisions from the Caucasus in the dead of winter, without adequate air cover. Army Group A managed to withdraw across hundreds of miles of steppe, while stiff-arming Russian forces pressing on its flanks in the south and east. At the same time, Red Army spearheads, converging from the north, were striving to close the Army Group's evacuation route by recapturing Rostov. The escape of the First Panzer Army through Rostov in January 1943, and the evacuation by sea of the Seventeenth Army from the Kuban Peninsula in October 1943, together deserve to rank above Dunkirk as an example of a beleaguered force withdrawing in good order against great odds.[14]

The saga of the Eleventh Panzer Division, commanded by General Hermann Balck (arguably the most talented divisional commander who fought in the Second World War) also deserves appreciative attention. It shows in particular why the student of military history inevitably embraces a tragic view of life. One sees outstanding generalship and the unflagging physical courage and tactical skill of the rank and file brought to nothing on account of poor strategic thinking. Hardly less important is that reading about such exploits reinvigorates one's capacity for admiration. Balck's achievements are an admixture of courage and high intelligence actuated by the pressure of events, as well as an illustration of a commander who was able to reconcile a keen regard for his troop's welfare with a demanding operational task.

On 7 December 1942, the Soviet Fifth Tank Army launched an attack across the Chir River, about ninety miles northwest of Stalingrad, and about one hundred miles north of Rostov, crashing through the Seventh *Luftwaffe* Field Division and the 336th Infantry Division, enroute to Collective State Farm 79, ten miles south of the river. The Eleventh Panzer Division had just arrived from the central sector to buttress General Manstein's expedition to relieve the Stalingrad garrison. Balck's panzers, outnumbered and outgunned, attacked the First Soviet Tank Corps at dawn on the 8 December; leaving in their wake the burning hulks of fifty-three T-34s. Over the next three weeks the Eleventh fought one engagement after another in the area between the Chir River and Rostov, often moving by night and attacking by day, as a means of catching the Russians off guard. Balck's division was largely responsible for destroying the Fifth Tank Army, even though the ratio of panzers to their Russian counterparts was never better than one-to-seven. The Russians lost some 700 tanks in their attempt to drive from the Chir to Rostov, which was a setback for the Red Army all the more demoralizing given that the opposing forces comprised a single, battered tank division, and three worn-out infantry divisions.[15]

Erhard Raus's Sixth Panzer Division fought with equal valor against obstacles of similar magnitude. Raus's account, along with complementary narratives by William Craig and Peter McCarthy, brings to the mind the inspirational value of history and can only help rehabilitate the idea of emulation, qualities that seem strangely exotic and irrelevant when reading our current strategic documents. In a manner similar to that of Balck, Raus's generalship highlights the difference between a leader (a term that has become hackneyed and destitute of meaning on account of its profligate misuse nowadays) and a mere manager discharging obligations imposed from above. Unaware of the scale of the gathering catastrophe at Stalingrad (he was even uncertain as to the exact destination of his division), Raus nevertheless prepared the Sixth Panzer for immediate combat as they entrained in France, contravening military practices and the procedures of the civilian railway authority. These regulations were sensible insofar as they were designed to economize the movement of large formations, but Raus knew that he could count on partisan attacks once the trains had crossed into Russia, and he suspected that as soon as the lead elements of his division arrived at the railhead they might well come under attack.

More conspicuous in Raus's memoir is the author's modesty and magnanimity, both enemies of the bureaucratic temperament. Most autobiographies written by general officers are built around cause-and-effect: the senior leadership took a given decision that either worked, or didn't. Fog, friction, and chance are hardly mentioned. Raus's commentary on the Stalingrad operation is the exception that proves the rule. His decisions merely provided the occasion for the resourcefulness and intrepidity of his staff, his subordinate commanders, and the rank and file. For example, Raus recalls that as his units were detraining in theatre, they immediately came under artillery fire, when suddenly, for no apparent reason, the enemy batteries fell silent at a moment when they might have been able to do serious damage. Raus explains that one of his regimental commanders, who happened to be at a nearby tank repair station when the barrage commenced, immediately ordered the maintenance troops onto such tanks as could still drive and led an attack on the artillery battery, annihilating it in quick time without suffering any losses. Raus's memoirs merely supplement any serious study of the Stalingrad campaign, but there is more insight here as to how things go than one might manage to extract from the *Quadrennial Defense Review* or *Transformation: A Strategic Approach*.[16]

Driving toward Stalingrad from the southwest, the Sixth Panzer Division, accompanied by the understrength Seventeenth and Twenty-third Panzer Divisions, managed to advance some sixty miles to within twenty miles of the enclosed Sixth Army. Along the way the force pushed through waves of counterattacking Russian tanks and infantry. The Soviet Second Guards Army, fully equipped with tanks and motorized infantry, stopped Raus's spearheads, but not before his division had liquidated a cavalry corps and a tank brigade. Dozens of T-34 carcasses dotted the steppe between the village of Pokhlebin and the Myshkova River.[17]

More prominent in reputation than either Raus or Balck is Erich von Manstein, whose operational achievements, such as Sickle Cut and the retaking of Kharkov in early 1943, go a long way toward justifying his renown. Antony Beevor notes that Manstein is today recognized as "the most brilliant strategist of the whole Second World War." Basil Liddell Hart, no doubt intentionally echoing Clausewitz's phrasing, described Manstein as a "military genius" who "combined modern ideas of mobility with a classical sense of maneuver, a mastery of technical detail and great driving power." Even so, Manstein's example shows that physical courage and intellectual ability do not always reflect moral stout-heartedness. History has assigned to General Friedrich von Paulus the blame for the liquidation of the Sixth Army, and there is a certain weight of justness in this verdict. But Manstein's own passionate justifications for decisions he made at the time are not entirely exculpatory. In his memoirs, he claims that on 19 December he—as commander of Army Group Don, which included the Sixth Army—ordered Paulus to break out; the "fact that the order did not achieve anything is due to the failure of Sixth Army Headquarters [i.e., Paulus] to carry it out." Manstein goes on to say that he chose not to threaten resignation, because to quit the field would have been an abdication of his duty to the forces trying to escape from the Caucasus, who might otherwise have been annihilated by the Soviet armies deployed around Stalingrad.[18]

A competing view is offered by Joachim Wieder, an intelligence officer assigned to VIII Army Corps, and one of the few survivors of the Soviet prison camps. In *Stalingrad: Memories & Reassessments*, Wieder argues that Manstein wavered not because he lacked an understanding of the fast-developing catastrophe, but from a species of moral feebleness. Wieder argues that Manstein should have immediately demanded adequate forces for the relief attempt—specifically, Army Group A, the mission of which was no longer practicable given the shattering of Army Group B—and that his resignation should have been ready in hand were Hitler to refuse. Wieder also points out that Hitler, who could not have forgotten his debt to Manstein for contriving the plan that yielded a spectacular victory over the French, would think twice about losing the one commander whose military judgments compensated for a lack of conspicuous National Socialist ardor. "Maybe such an unbending stance in the interests of the whole, would have served to force the pig-headed Supreme Commander of the Armed Forces to give in to crucial decisions," Wieder states, "he being particularly dependent on the ability of his Field Marshal at this critical time":

> Von Manstein did not do this. Thereby he not only lost complete sovereignty on his home ground of strategy, he also lost an essential part of his personality. He bowed to the pressure of obeying despite superior insight, and resigned himself to becoming the executive organ of a dilettante leadership that refused to recognize the foreseeable. With this, he took on a far greater measure of responsibility than did his unhappy subordinate General Paulus, because he had the broader overview and saw more clearly what the game was and what needed to be done.[19]

Compelling though it is, Wieder's memoir is not the last word on the subject—if only because there can be no last word on any catastrophe brought on by an admixture of oxlike persistence, bigotry, stupidity, and moral frailty set against astonishing examples of courage, self-sacrifice, and rock-of-Gibraltar faith, sometimes frightfully misbegotten, and sometimes richly justified, in one's commanders, countrymen, the aims of the war, and oneself. Herein lay the import of studying history, and the limited worth of obsessing over technological matters.

What matters to the modern officer is not the clash of weaponry or the deployment of arrows on a map—which are worthy, if subordinate, fields of study—but the interaction of character and circumstance. The moral and critical intelligence of the modern-day commander can only benefit from pondering the issues that the Stalingrad campaign sets before him. What was possible at the time? If faced with circumstances that are similar in the abstract, what are the likely choices, and which is the most prudent? In preparing to face the pressure of events, how does one reconcile the obligations of career and professional morality? Studying this campaign or any other with suitable intensity can bestow a capacity for judgment that cannot be realized in any other way.

Friedrich von Paulus is little esteemed in the eyes of posterity, largely because he superintended a defeat—entailed by suffering that remains astonishing, even by the lights of a century that was promiscuously violent—that he might easily have forestalled or at least mitigated. Even so, a sympathetic consideration of his action and character is not without value to our own times, particularly our approach to developing commanders and staff officers. The tragic circumstances of von Paulus's professional life might point the way toward a rehabilitation of our own understanding of how we select officers for promotion, and also how we determine a suitable progression of assignments. Nowadays we attach great weight to breadth of experience that can only come at a cost of developing genuine resourcefulness. By means that resemble industrial processes, we work toward creating a mass-produced tool and not a mind, while the aptitude of the individual officer is discounted.

We strive to produce an officer corps that is homogeneous in its habits of mind. The greater the rank the more stringent the uniformity of outlook. For example, every senior field commander must attend at least one, and usually two or three, in-residence professional education programs in an ordinary career, irrespective of his or her ability to perform in a demanding academic environment, with the predictable consequence that these programs are gutted of substance. Staying too long in one operational assignment, or a series of operational assignments, and eschewing in-residence professional schooling is as mortal to one's career prospects as moving from one staff job to another midcareer, even if this might be in the best interests of military efficiency. The endorsement of one's commander on a performance report (the "push line") cannot read "excellent staff officer, ready for more demanding staff work" if it is meant to advance the officer's professional prospects.

It would be wise for us to acknowledge that, at the highest levels, command and staff work are distinct vocations, and that it is nearly impossible to unite in one person the boldness of the commander and the intellection of the staff officer. We should consider developing separate career paths for the gifted staff officer, even as we should not undermine the professional development of the talented commander by forcing him to spend time in academic programs and ticket-punching staff assignments for which he is not suited. Personnel management priorities, moreover, should be built around pairing these complementary experiences and temperaments in the form of commander/chief of staff, rather than forcing officers into a procrustean template that might gratify the bureaucratic instinct, but which is at odds with reason and experience.

One unfortunate result of the status quo is a culture all too eager to justify itself independent of critical reasoning. It is not unusual to hear from senior officers the following apologia: "Serving in a mixed multitude of assignments has helped me gain a broad perspective and approach the summit of my profession." Leaving aside the self-flattering and panglossian sentiment in which remarks of this kind are bathed, such a point of view is unhindered by the recognition that high rank, by itself, is a sign of administrative responsibility and does not necessarily reflect wisdom, if only because promotion boards strive to create a cadre of senior officers untroubled by unsafe ideas. A product of our current system knows well how to transact ordinary business rapidly and efficiently, but independent judgment is rarely tested, if only because a philosophical difference with one's boss at any time can terminate the possibility of advancement.

It is on this subject that Friedrich von Paulus, which posterity has made synonymous with military disaster, bears contemporary significance. His career is relevant to us insofar as it sheds light on the weaknesses in our current understanding of how best to manage the great talent of our officer corps. There are no ready-made alternatives advocated here, only a suggestion that we ought to consider revamping our assignment and promotion policies, and this must start with looking at officers as intellectual rather than administrative capital.

Paulus was a first-rate staff officer who was pushed into senior command, a task for which he was ill-suited, largely because he was what we infelicitously call nowadays a "team player." Today, our difficulty is of the same character, but the circumstances are reversed. We have no shortage of superior field commanders; it is in the arena of strategic planning that we might benefit from recasting how we educate, train, and promote officers. An effective way to address this is to make sure that we cultivate the sort of intellectual excellence that strategic planning requires, while at the same time allowing those officers with an aptitude for command the chance to develop such gifts without having to spend time in cubicle farms at staff headquarters, or attending a succession of in-residence education programs that benefit neither the officer nor the college.

By almost any measure Friedrich von Paulus was a first-rate staff officer who was by training and temperament poorly suited for leading massed formations in combat. At the age of nineteen he abandoned a budding career in law for the profession of arms. His initial choice of occupation is significant. One succeeds at the practice of law not by acting on instinct, but by evaluating precedent and the facts at hand with deliberation, under the cold, bright light of reason. Boldness and perseverance have their place—timidity and a weak will are punished in the practice of law as they are in almost any other profession—but only as the pinnacle of a diligently constructed edifice. Paulus's initial choice of profession thus foretokens his aptitude for staff duty: an intellect drawn to ponder abstractions, fastidiousness in matters of detail, and a tendency to approach a problem by considering all dimensions and angles before taking action.

Commissioned an army officer in 1911 (the navy turned him down on account of his modest social background), Paulus served with distinction at Verdun and elsewhere, and earned the Iron Cross, Class II and I. Most of his experience during World War I, however, was as a staff officer widely respected for his sedulousness, mental agility, and punctilious manner. Commanders at the General Staff and regimental level unreservedly trusted his judgment—the highest compliment, given the stakes of World War I.

Following the war, Paulus remained in the army and moved up in rank, chiefly because of his superior staff skills. Walter Goerlitz's biography of Paulus includes an illuminating extract from a fitness report on him, when he was serving in the early 1920s as adjutant (chief of staff) of the Fourteenth Infantry Regiment. An "exceptionally good and enthusiastic soldier," the report states. Captain Paulus is:

> methodical in his office work and is passionately interested in war games, both on the map and the sand model. In these latter he displays marked tactical ability, though he is inclined to spend over-much time on his appreciation, before issuing orders.[20]

During his tour of duty as an instructor at the General Staff College, Paulus's phlegmatic temper earned him the nickname "Fabius Cunctator," after the Roman general known for prudence and circumspection, rather than impulsiveness or passion. General Ferdinand Heim, whom Hitler had court-martialed (he was later exonerated) because his Forty-eighth Panzer Corps failed to engage Red Army forces working to encircle the Sixth Army at Stalingrad (Heim was closing in on the Russians when Hitler ordered him to attack in a different direction) recalls that from the beginning, Paulus's command prospects were hardly spectacular. During a training exercise the commander of the Fourteenth Infantry Regiment was "killed," leaving the adjutant, Captain Paulus, in charge. The evaluator noted that Paulus "lacked decisiveness."

Even so, it would be a mistake to conclude that Paulus (as is the case for any talented staff officer) was constitutionally unable to command; most of the time he served reasonably well when called upon to do so. During the mid1920s, he and Erwin Rommel held company commands in the Thirteenth Infantry Regiment. Paulus's unit, Walter Goerlitz points out, "invariably beat [Rommel's machine gun company] at both work and play. A General Staff Officer," Goerlitz adds, "is not always and of necessity a rotten Company Commander!"[21]

In 1934 Lieutenant Colonel Paulus briefly commanded a motor transport unit, which would be the last time he would command until 1942, when he took charge of the Sixth Army. Under his leadership the motor transport unit was reconstituted as an armored cavalry formation—the prototype of the motorized infantry divisions upon which blitzkrieg would ultimately depend. Promoted to Colonel in 1935, Paulus replaced Colonel Heinz Guderian as the chief of staff of Headquarters, Mechanized Forces, in Berlin—the outfit responsible for the development of the panzer and mechanized infantry divisions. Guderian had served in the position for about fourteen months before moving on to command the Second Panzer Division, and not surprisingly, given his swashbuckling manner, harbored misgivings as to Paulus's fitness for the position. But as Walter Goerlitz points out, Guderian's "doubts were soon allayed."

> Paulus in his eyes quickly became the personification of the wise, enterprising, enthusiastic and conscientious General Staff Officer of the modern type, whose sincerity was beyond all doubt a pearl of great price.[22]

Paulus's specific contributions to the development of the panzer arm are not known. He cared not at all for flamboyant self-promotion (he took to heart the idea that staff officers "have no names"), but his work must have been of a high standard if his promotion to deputy chief of the General Staff in 1940 is any indication.

In this post, Major General Paulus superintended the writing of Operation *Barbarossa*, the plan for the invasion of the Soviet Union. His understanding of Germany's strategic situation remains unheralded, because it was never given extensive consideration. Paulus advanced the idea that Germany should focus on the Middle East due to its abundant supplies of oil. Control of this region would knock Great Britain out of the war and solve Germany's acute shortage of petroleum at the same time. The Soviet Union could then be dealt with in detail. As perceptive as his idea was, Paulus—the embodiment of the apolitical officer—did not clamor on behalf of this strategy, and preferred to leave such decisions to the head of state and his closest advisers.[23]

In January 1942 Paulus was promoted to full general and given command of the Sixth Army. The appointment came as a surprise; Paulus had never commanded a regiment, and there were many officers senior to him who were more suitable choices. Paulus was promoted based not on his fitness to command, but because of the influence of what nowadays would be termed his sponsors. Hitler acted on the recommendations of General von Reichenau and General Halder. Paulus had served under Reichenau as chief of staff, Sixth Army, 1939–1940. Starting in late 1940, he worked for Halder as deputy chief of the General Staff. Both men appreciated Paulus's keen mind and energy, but above all they valued his unyielding loyalty.

Disaster did not long follow, though at first Paulus acted with his customary efficiency and perceptiveness. In late June 1942 the Sixth Army began its move east toward Stalingrad, and it was here that Paulus's gifts as a staff officer came into play. Such battles as were fought resembled more closely the antiseptic war games of the staff college. Clausewitzian fog and friction remained at manageable levels. Paulus's greatest concerns involved the administrative elements of command: making sure his troops had adequate stocks of fuel, ammunition, and other necessities. Meanwhile, the Sixth Army piled up what Count von Schlieffen identified as "ordinary victories." The enemy simply vacated the battlefield and so managed to avoid a trial of strength that almost certainly would have gone in the *Wehrmacht's* favor.

In September, with the Sixth Army heavily engaged in the fight for Stalingrad, Paulus expressed concern about his insufficiently protected flanks. Hitler listened sympathetically, but dismissed the matter based on intelligence estimates that the Red Army had no reserves to exploit the situation. In mid-November, reports filtered up to Paulus that the Russians were massing forces at points northwest of the city, about sixty miles distant, where the Red Army held both banks of the Don River. Meanwhile, in Stalingrad the intense fighting of the previous two months had given way to sniper activity and the occasional skirmish. Both combatants were exhausted, but the strategic initiative had passed to the Red Army. Acting on Paulus's repeatedly expressed concerns about the flanks, Hitler moved the severely understrength Forty-eighth Panzer Corps into a backstop position northwest of Stalingrad, a sector held by poorly trained, poorly equipped, satellite armies. It is significant that neither Paulus nor his immediate superior, General von Weichs, head of Army Group B, had control over the Forty-eighth, the Army Group's only armored reserve, which was a circumstance that should have prompted Paulus to create his own.

It is on this point that the staff officer's frame of reference begets disaster. Even though it was his own troops that were jeopardized by the Red Army build-up, Paulus did nothing because the threatened sectors were the responsibility of Army Group B. The Sixth Army chief dutifully moved ominous intelligence reports up the chain of command, assuming that higher headquarters would act appropriately. A battle-tested commander would have maintained a full mobile reserve ready for the inevitable surprise that war, at the operational level, almost always delivers. He also would also have adjusted his logistical system to support his reserves. Paulus never bothered to consider these things; the panzer formations under his command were being chewed up in street fighting, while most of his supplies were kept in depots on the west side of the Don. Rather than react on his own initiative to the incipient danger, Paulus unthinkingly followed Hitler's orders to eject the Russians from Stalingrad. Paulus, the ever-loyal staff officer, did as he was told, and nothing more.

The upshot here is that the regulated tempo of staff work cannot prepare an officer adequately for the fermenting chaos of operational combat. An experienced commander knows when to obey orders to the letter, and when merely to gesture in that direction and otherwise adjust to immediate circumstances, an outlook consecrated by time in military culture. The commander in the field senses—by an informed intuition—gathering peril and acts to forestall or subdue it. He can read the morale of his subordinates, has firsthand access to prisoner interrogation reports and the like, and thus can judge and shape the trajectory of battle in ways that higher headquarters personnel, hundreds or perhaps thousands of miles away, cannot. The expressions on the faces of frontline soldiers, the condition of the equipment, the piles of expended shells, the casting of a tactical engagement, and the scenes of battle are all interpreted by the lights of the commander's temper, which must consist of a capacity for boldness that may strike the staff officer as recklessness. The field commander's wisdom is purely utilitarian; it is judgment based on a compound of experience, good sense, and above all else, information available at the moment. Right conduct is based on quickly sizing up the range of possible actions and their immediate consequences, and fashioning victory out of the raw material of battlefield fog and friction. This is the essence of command—an exotic and inscrutable world to the officer bred to staff work, just as the staid precincts of staff duty are alien to the officer who is at heart a field commander, the officer who prefers to react rather than to ruminate over comprehensive sets of information.

Paulus's intellectual and professional gifts were of no use to him or the army he commanded once the Red Army sealed the pocket. The situation demanded prompt, resolute action; every minute that passed weakened his army, even as the ring around the Sixth Army became stronger and their would-be rescuers were pushed farther away. Message traffic between Paulus, Weichs, and Hitler proves that Paulus was aware of what needed to be done, but he simply fell back on his staff-officer reflexes, advising his chain of command, and waiting for them to make the call. The situation deteriorated rapidly. The first starvation cases were reported on 9 December 1942, about the same time as it became obvious that the airlift to provision the army could not possibly succeed. Paulus did nothing. When relief forces approached to within twenty miles of the pocket he refused to act, claiming that his panzers had enough fuel to move only twelve miles. Hitler used this information to justify his decision to keep the Sixth Army in place.

Knowing Hitler's mind as well as any other senior officer at the time, Paulus might have broken through to Raus's spearhead and then presented Hitler with a fait accompli. Alternatively, he might have inflated his fuel figures or pressed his logisticians (by nature quartermasters are cautious) to scrounge and recalculate, thus goading Hitler into making the right decision. But Paulus apparently never considered doing so. Within six weeks the final butcher's bill had been paid: the dead, wounded, and missing numbering in the hundreds of thousands.

Paulus has been faulted on account of his weakness of will and confusion, not illegitimate judgments, but the source of the problem lay in his unsuitability for a field command. His promotion was an act of patronage: a favored subordinate was moved ahead as a reward, with disastrous consequences, something that should prompt us to rethink our own approach to officer professional development.

Like all epic battles, Stalingrad defies summary, and will overawe any commentary that professes to speak with exhaustive authority on the subject. With each new generation military historians will turn to the battle as a means of understanding not only the past, but the present as well. For us, the struggle is an unforgettable portrayal of the primacy of intelligence and moral character in war, which we ignore or downplay at our own peril. A study of this campaign also offers us a salutary rebuke in regard to how we identify and groom officers for senior positions, by emphasizing technical specialization and uniformity of outlook at the expense of authentic resourcefulness—an unwieldy appendage of our industrial approach to waging war. In particular, we ought to acknowledge and advance those officers who demonstrate the right conceptual framework for staff work, which at its highest pitch differs from those habits of mind that make for outstanding command in the field.

Chapter 6

Desert War: From El Alamein to Cape Bon

The portion of the North African littoral that stretches from the Mediterranean Sea to about forty miles inland between Bizerta and Port Said was an ideal battleground for European combatants equipped with up-to-date weaponry. Operations were not impeded by major population centers. There was no refugee problem, nor did the prospect of irregulars dressed in civilian clothes aggravate operational difficulties or provide the occasion, and often the excuse, for murder, as was the case in Russia, and later in France (Oradour-sur-Glane comes immediately to mind). The inhabitants of the few settlements along the North African coast had no compelling stake in the outcome of the struggle between Axis and Allied forces; they tended to treat the various armies that passed through the region either grudgingly or generously, depending on individual experience (though the colonizing Italians were widely disliked).

The geography and climate, though often hostile to man and machine, facilitated more than hindered military operations. The clear skies were a boon for pilots used to the weather patterns of northern Europe. Tank crews found that the coastal tableland (mostly limestone dressed with a layer of fine sand, clay dust, and pebbles) made for good going. Even the soft-skinned vehicles could drive across country for considerable distances. The British Long Range Desert Group, equipped with Willys Jeeps and Chevrolet 30-centoweight trucks, moved freely across hundreds of miles of trackless desert to conduct raids deep behind Axis lines—a span of maneuver that could not be equaled in Europe for large portions of the year, if at all. No army can live off the desert, but in the mechanized age water, fuel, ammunition, troops, and equipment can be transported to and from Europe via the handful of ports, and from there trucked to the front along the Via Balbia highway. The ratio of force to space was quite small, so operational maneuvers gave free play to the skill, resolution, and perceptiveness of the combatants.

Scarcely less important is that the battle for North Africa involved relatively small forces, held a low strategic priority, and was far enough removed from Berlin and London so as to attenuate the ideological animosities and titanic ambitions that envenomed war on the continent. The North African campaign thus took on the feel of a private war between gentlemen beholden to traditional military courtesies. The near complete absence of atrocities put the campaign at the polar opposite from the war waged in Russia, and gave it an almost anachronistic tinge. Front line troops would on occasion grant each other informal truces. "Sportsmanship showed on both sides," an Afrika Korps veteran recalls. "Football games were not interrupted by artillery fire during certain periods." Rommel's reputation was built on his chivalric conduct as much as it rested on his dash and valor. During the battle of Gazala (one of Rommel's more spectacular victories) a captured British officer complained on behalf of his fellow prisoners about the scanty allocation of water. In response, Rommel declared that he and his soldiers lived on the same ration, one-half cup a day, and that if the supply situation didn't soon improve he would have to ask General Ritchie for terms. Montgomery and his soldiers acted very much the same way. Montgomery and Rommel both savored the opportunity to meet captured senior officers, invigorating the increasingly dormant idea of chivalry that was one of the many civilized bequests of the Middle Ages. The oft-published photograph of Montgomery's meeting with General von Thoma shortly after the latter surrendered is a neat expression of the temper of the desert war. How unlike the Eastern Front, where Russian prisoners by the thousands were starved to death, and where captured German troops were murdered, their corpses horribly and pointlessly mutilated![1]

Not that there weren't occasions when war crimes might easily have stained the honor of the combatants. A communiqué from Berlin after the fall of Bir Hakim declared that French prisoners would be executed as partisans. De Gaulle immediately broadcast over the BBC that "with profound regret" he would, in turn, have no choice but to "inflict the same fate" on German prisoners. Within hours Berlin made the following announcement: "On the subject of the French forces who have been captured in the fighting at Bir Hacheim, no misunderstanding is possible. General de Gaulle's men will be treated as soldiers." During the summer battles of 1942, Rommel learned that

British headquarters had authorized the deprivation of food and sleep of German prisoners as a means of extracting information. Rommel, over a clear radio frequency, protested, and noted that the treatment would be reciprocated. Shortly thereafter the British, also using unencoded radio transmissions, revoked the order. This is more than merely a fastidious adherence to the spirit of the Geneva Conventions. These incidents show that commanders can indeed mitigate the native brutality of war, and that war does not inevitably degenerate into savagery. The North African campaign had its share of killing festivals. The gruesome reality of tanks crews immolated when an antitank round ignited ammunition bins, or infantry units cut to pieces by artillery fire, was very much in evidence in the desert war. Both sides suffered high casualties, especially during the El Alamein battles and the investment of Tobruk. Even so, the conduct of the combatants shows the extent to which the moral intelligence of commanders determines the cast of war. For all of its reprehensible features, war also brings forward magnanimity and selfless regard for human dignity that is all but nonexistent in more humdrum kinds of social intercourse, such as business and domestic politics.[2]

Nowadays, the desert war remains of interest to military strategists and historians for a couple of obvious reasons. The advanced weaponry employed in North Africa is a subject in tune with contemporary habits of mind. The competition between the Germans and Allies to produce a better tank was at its sharpest. The Germans did not face an enemy equipped with a tank greatly superior to their own (as they had in Russia in 1941), even though the heavily armored Matilda tanks and the American-made M3 Grant caused their fair share of trouble. The Flak 18/36 guns, first used as an antitank weapon by the Condor Legion during the Spanish Civil War, decimated allied tank formations and demoralized those crews who had not experienced firsthand the hardhitting 8.8cm armor-piercing shells. The appearance of the long-barreled Mark III and IV tanks tilted tactical engagements in favor of the Germans in ways not possible on the Eastern Front. On the other side of the hill, the arrival of the M4 Sherman tank in late 1942 moved matters back in favor of the Allies. The employment of the new Tiger I tank—basically, the dreaded "88" set in a fully rotating casemate mounted on tracks—shortly after the commencement of Operation *Torch* moved the technical balance back in favor of the Axis forces, though their small number (only about three dozen were employed) meant that they were only an occasional nuisance for British and American forces. Even so, the Tiger I was fearsome. On 20 March 1943, a half dozen Tigers, without a single loss to themselves, destroyed thirty-five Shermans west of Maknassy Pass, and stopped the U.S. Ninth Armored Division from reaching the sea. As would be the case in Europe, most Tigers in North Africa were destroyed by their own crews on account of mechanical failure, or because they ran out of fuel.[3]

Equally of interest for the modern student of war is the interdependence of the various arms. Naval, air, and ground forces benefited from, and so came to rely on, each other's particular strengths, more so than in any other previous campaign, which is hardly surprising given that the North African operation was for all the combatants an expeditionary operation.

Today's campaign planner might well look upon the desert war—from the German perspective especially—as a validation of our current priorities. In matters of technology and military skill, our forces today are every bit the equal of, if not superior to, the battle-tested Afrika Korps. We have solved both of the problems (logistical inadequacies and the absence of air superiority) that bedeviled General Rommel. Indeed, the U.S. Navy and the USAF stake their budgetary claims not only on making sure that the skies and seas are clear of the enemy's presence, but also in providing Joint Force commanders with operational reach, that is, the ability to deliver troops, material, and firepower unhampered by the twin tyrants of geography and time. At the strategic level, a map of the region makes plain to even the amateur strategist that Malta needed to be secured in order to safeguard lines of communication before placing a force of any size in North Africa. We would not make a similar mistake today.

It is otherwise tempting to dwell on the military legends and exotic features of the North African campaign, particularly the exploits of Rommel and the idiosyncrasies of Montgomery, with his plethora of flamboyant hats, and ignore or downplay its timeless qualities, which are quite substantial. There is much underappreciated treasure here, not the least of which is the extent to which the skill and ambition of individual commanders—fighting what was widely assumed to be a marginally significant series of battles—altered the strategic direction and outcome of the entire war. The imposing stamp of individual ability is something we ought to contemplate as we ponder our contemporary approach to officership and military planning. For all of its modern characteristics, the North African campaign reminds us that war is not a "process" that can be broken to the saddle of managerial savvy and technical sophistication.

It's worth considering for a moment the strategic calculations that initially gave occasion to the war in North Africa. Envious of Hitler's conquests, Benito Mussolini hoped to expand his colonies in Libya into a latter-day Roman Empire, and this required the elimination of British hegemony in the Mediterranean basin. In late 1940—a few months after declaring war on Great Britain and France (in the midst of her death throes)—Mussolini ordered his armies to invade Egypt. Brushing aside a thin screen of British troops, Italian forces crossed the Egyptian frontier and soon came within striking distance of Alexandria. Not far beyond lay the Nile Delta and the Suez Canal, the crucial artery connecting Great Britain with its colonial possessions in the Middle East and the Indian Ocean.

For all of his contemptible braggadocio, Mussolini's attack was marvelously opportune. In the autumn of 1940, the British, still licking their wounds in the aftermath of Dunkirk, were otherwise absorbed in fighting the Battle of Britain. The advance of Mussolini's armies posed a menace not quite as immediate as that of Hitler's legions awaiting the order to invade England, but the threat was indeed compelling given the paucity of His Majesty's forces in North Africa. For Great Britain's weakness was not her proximity to German forces in France, but her dependence on free use of the seas. To lose control of the Mediterranean would mean that strategic materials—oil most especially—must move round the Cape of Good Hope, and this would expose to a much greater degree her merchant fleet to German submarines and surface raiders. And even if these threats were controlled, the longer sea routes would reduce the flow of vital supplies to a strangulating trickle.

In response to the Italian aggression, a small British force under the command of General Richard O'Connor struck back in December 1940. Though greatly outnumbered, O'Connor's troops smashed the Italian armies, and—propelled largely by captured vehicles, fuel, and food—drove them back to El Agheila on the border of Tripolitania. The culminating battle took place near Beda Fomm, an operation of encirclement and annihilation worthy of any the Germans had pulled off. British casualties were negligible. Fewer than 2,000 troops were killed, wounded, and missing, which was a meager price to pay for an advance of 500 miles and the capture of 130,000 Italian soldiers, some 400 tanks, and nearly 1,000 guns. On 9 February 1941—three days before Rommel and his advance party arrived at the Castel Benito airport in the Libyan capital—British troops were poised to advance on Tripoli, which lay about 500 miles to west, across the largely undefended Via Balbia highway.[4]

Unlike Italy and Great Britain, Germany in February 1941 had no strategic interests in the Mediterranean region. With the commencement of Operation *Barbarossa* only a few months away, and campaigns against the Scandinavian countries and the Balkan states emerging on the immediate horizon, Germany had little offensive power to spare for what could only be an excursion. Even so, the collapse of the Italian army was not without danger. With the British in command of the Mediterranean, the southern flank of the Third Reich would be vulnerable to counteroffensives from Sardinia in the west, all the way to Crete. Even more, the weakening of the Axis presence in the Mediterranean would justify the decision taken by two pivotal neutrals—Spain and Turkey—not to collaborate with Nazi Germany in prosecuting the war.

Thus, Hitler decided to provide a small mobile force, one panzer and one motorized division, to help the Italians restore the situation, and collaterally, tie down British forces outside of Europe. Unusual for Hitler, success in the African theater was initially defined as stalemate rather than conquest, a point exemplified by the chain of command. Sensitive to Italian concerns that Germany would domineer over her militarily ineffectual ally, Hitler made Rommel subordinate to General Garibaldi, the chief of staff of the Italian armed forces. The German High Command also instructed Rommel to refrain from major offensive operations. Quite unintentionally, the political arrangements and the relatively diminutive size of the Afrika Korps led Rommel to demonstrate to the world something that the Germans would soon perfect in Russia: the realization of the Clausewitzian idea that offensive action is an essential component of a successful defense.

What happened over the next two years has been the subject of a multitude of books. The struggle for North Africa took on the character of a reciprocating machine. Rommel would attack British forces, who would fall back—sometimes in great disorder—hundreds of miles to the east. His supplies exhausted, Rommel's armies would pause, and then recoil from the momentum of a British counteroffensive. Rommel would counterattack, the British would withdraw and rehabilitate themselves (thanks to a shortened supply line), and so on. The campaign swung permanently against Rommel in July 1942 at the first battle of El Alamein, in which his last serious effort to reach the Suez Canal broke apart against stiff British resistance at Ruweisat Ridge. In October 1942 Montgomery's Eighth Army went on the offensive and drove the Afrika Korps westward. Mersa Matruh, Tobruk, El Agheila, and a clutch of other sites that had been occupied by the British, then the Germans, and then the British again (Benghazi changed hands five times) over the preceding months, all yielded for a final time to the advancing British army. In the meantime, Operation *Torch*—the Anglo-American invasion of northwestern Africa—gathered momentum. By spring 1943 the Afrika Korps, despite generous reinforcements from Germany, was crammed into a shrinking perimeter bounded by Bizerta and Enfidaville in Tunisia. On 13 May, about ten weeks after Rommel departed for a hospital in Europe, the shooting had come to an end. Nearly a quarter of a million Axis troops entered captivity.

The aftermath was largely the opposite of what either side might have foretold in 1940, demonstrating the extent to which the consequences of battle ramify in ways that the combatants can scarcely be expected to fathom. Instead of acquiring an empire, Mussolini was deposed, and in late 1943 Italy joined the Allies. By then Italy had become a forward garrison for German troops—the very thing Mussolini and Hitler hoped to avoid. The Germans could take arid consolation from the myth of Rommel's military genius, but strategically they were in a bad way. The southern flank of the Third Reich was now open to assault across a broad front, not only because the men best suited to rebuff an invasion of Italy were dead or crammed within POW cages, and their equipment destroyed, but also because the Allies had gained vital experience in amphibious operations. On the other hand, the Axis war effort benefited, perhaps, by the extent to which the North African campaign absorbed allied resources, most especially time.

The British achieved what they had set out to do by keeping the Axis from cutting their lines of communication with the empire, but at a cost in men and matériel that forced them to yield leadership of the Western Allies to the United States sooner than anyone might have expected. This development aggravated an existing, if latent, animosity between the Americans and the British. Several U.S. generals left North Africa with their Anglophobia confirmed, among them Patton, Clark, and Bradley. There was no love lost toward the Americans from the British senior commanders, including Air Marshal Tedder, General Alexander, and most especially General Bernard Montgomery. Most British officers were discreet on this point, but Montgomery's petty, self-injurious arrogance would cause serious problems in northwest Europe later on—yet one more example of the friction that inevitably attends coalition warfare, even when the partners are bound by deeply rooted cultural ties. From 1943 onward, a muted but unrelenting internecine struggle played out between the British and the Americans in regard to strategic and operational planning. Had it not been for General Eisenhower's exceptional diplomatic skills the war might have dragged on longer than it did. This is a point worth remembering as we contemplate relying heavily on allies as part of our national military strategy.[5]

Subordinate British commanders were hardly free of the same vice, up to the point of trafficking in sham stories that are easily uncovered. Lieutenant General Brian Horrocks, who commanded the British XIII Corps at El Alamein, indulges in his memoir the fiction that the Sherman tank was a binational creation. "I feel the time has come for the history of the Sherman tank to be made public," Horrocks declares in his book published fifteen years after the war ended. Incredibly, he argues that in 1941 the Sherman was funded by the British and built to their specifications. "At the time it was not considered diplomatic for us to claim that the improved fighting qualities of the Sherman were due to British brains," Horrocks declares. "The Sherman was a really first-class example of Anglo-U.S. cooperation, the fighting part British and the mechanical side American." This nonsense springs from the self-flattering idea nurtured by British generals that they were great soldiers, and their U.S. ally very able technicians, but at war clumsy amateurs.

The truth is that the United States at that time did not have a weapons industry that catered to the needs of foreign countries. The Sherman was built to U.S. Army specifications set down in March 1941 by the Armored Force Board—before Rommel and his Flak 18/36 gun crews helped the British understand the multifarious inadequacies of their own tank designs. The Sherman's main armament, the M3 7.5 cm cannon, fired both armor-piercing and high-explosive shells (shells contain an explosive charge that is detonated by a fuse). By contrast, the British army was the only major combatant in the Second World War to rely on shot, which does damage strictly by its mass once it penetrates armor. Neither of Britain's main tank weapons, the two-pounder (approximately 4.0 cm) and its successor the six-pounder (approximately 5.7 cm), could fire high-explosive shells. In late 1942 the British armaments industry copied the M3 cannon in order to address what had proved to be a grave deficiency.[6] It is hard to believe that Horrocks, or any informed reader of his memoirs, would not have known as much. The only explanation for Horrocks's remarks is national chauvinism utterly impervious to fact.

The irony of the North African campaign is that both the Germans and the British adopted defensive strategies in order to keep the theater from becoming a major draw on forces and equipment needed elsewhere. But the steps that both sides took, independently of each other, brought about the very circumstances they had hoped to forestall. After Beda Fomm, in early 1941, the British were on the verge of throwing the Italians out of North Africa. General O'Connor very much wanted to press on to Tripoli, but his boss, General Sir Archibald Wavell, commander-in-chief, Middle East, thought otherwise. With the Italian threat neutralized, General Wavell turned his attention to the gathering Nazi menace in the Balkans, a view shared by his political chief in London. On 12 February 1941, the day Rommel arrived in Tripoli, Winston Churchill signaled Wavell:

> We are delighted that you have got this prize three weeks ahead of expectation, but this does not alter, indeed it rather confirms, our previous directive, namely that your major effort must now be to aid Greece and/or Turkey.... Our first thoughts must be for our ally Greece, who is actually

fighting so well. . . . Therefore it would seem that we should try to get in a position to offer the Greeks the transfer to Greece of the fighting portion of the Army which has hitherto defended Egypt, and make every plan for sending and reinforcing it to the limit with men and material.[7]

British maritime and land forces were transferred to Greece or, in the case of the Seventh Armored Division, sent back to Egypt to rest and refit. The headquarters staff of the XIII Corps was broken up and with it went the expertise that made the lighting victories over the Italians possible.

In a sentiment that Rommel would have approved, General O'Connor in retrospect believed that he should have immediately attacked, making the withdrawal of his forces impracticable. The British might easily have taken Tripoli on the run and still been reconstituted in time to meet the German invasion of Greece, which did not take place until April 1941, a month after the forces taken from North Africa had arrived. The British attempt to spare Greece failed miserably, while the Axis position in North Africa was transformed by the arrival of the Afrika Korps under the command of General Erwin Rommel—an example as good as any other of the uncertainties of war.

By the lights of our current system-of-systems approach to strategic planning, Rommel was beaten the moment he arrived in theater. For starters, British forces outnumbered the Afrika Korps. The British chain of command, moreover, went from London to theater commander to army commander. By contrast, Rommel nominally worked under the Italians and depended on them for his supplies, but his true military masters resided in Germany. And thanks to British decoders in London, orders sent from Berlin to North Africa were rarely kept confidential. Rommel turned these liabilities into advantages. Rommel used the bifurcated command structure in much the same way as an adolescent plays one parent off against the other; he would go directly to Hitler for approval of a plan that he knew the Italian chief of staff would reject. If necessary, Rommel would go the other way and justify a decision that the General Staff disliked on the grounds that he had squared matters with the Italians.

By contrast, his opponents were beholden to Churchill's wishes, something Rommel manipulated in his favor by piling up one lightning victory after another. The more conspicuous Rommel's success, the more friction he managed to inject between the British army commanders and their political chief. In fact, Rommel's most important victories were his earliest, because they utterly astonished the British. Based on intelligence gathered from Ultra, the British decoding apparatus, General Wavell was convinced, and London confirmed this, that the Afrika Korps could not go on the offensive until May 1941, April at the absolute earliest. Yet Rommel defied direction from Berlin and attacked immediately and with breathtaking effect, thus founding the Rommel mystique of invincibility.

Rommel had fewer inhibitions than O'Connor about ignoring the dictates of his commanders, and this raises the question as to why he was assigned to lead the Afrika Korps in the first place. Based on his sterling record as an instructor at the Potsdam military academy, Rommel was chosen to command Hitler's security battalion during the occupation of the Sudetenland. Hitler thus came to know Rommel personally. Rommel was untainted by what Hitler viewed as the General Staff clique, moreover, and Hitler undoubtedly saw in Rommel a man of a similar cast: a risk-taker whose decisions were based on intuition rather than befuddling calculations.

By any other reckoning the appointment was a mistake. Rommel was ill-suited to conduct holding operations, and was hardly more amenable to following orders if he thought more might be accomplished by disobedience. These traits are abundantly in evidence in his memoir of the Great War, *Infantry Attacks!* (1937). His experiences in France made him even more headstrong. His Seventh Panzer Division had raced across northern France and, having overcome Lord Gort's counterattack at Arras, wheeled south to capture Cherbourg, along with 30,000 French prisoners. His was the first division to reach the English Channel. The unpredictability and elusiveness of the Seventh Panzer earned it the nickname, "The Ghost Division." Cautious skirmishing was for him a far greater operational risk than pushing ahead with exposed flanks. All of this was part and parcel of his aggressive, ambitious temperament. And it was this temperament more than any grand strategy that determined the trajectory of the North African campaign.

Though under orders to conduct a defensive operation, Rommel went on the attack as soon as he had what he judged to be the minimum forces necessary to do so. The Afrika Korps gained territory for the Axis powers, and more importantly, discomfited not merely opposing generals, but the British head of state as well. In this way Rommel not only mastered the battlefield, but also forced the enemy to alter his strategic priorities. Embarrassed by Rommel's successes, and desperate for a victory of any kind, if only to fortify the morale of British citizens and soldiers alike. Churchill poured reinforcements into North Africa and pressured his theater commanders to go on the offensive. When their essentially prudent judgment counseled otherwise, Churchill sacked them. Wavell was replaced by General Sir Claude Auchinleck (June 1941), who in turn was replaced by General Sir Harold Alexander (August 1942). Field commanders were not spared. O'Connor, who was captured by the Germans in April 1941, was succeeded by General Sir Alan Cunningham, who was subsequently replaced by General Sir Neil Ritchie as commander of the Eighth Army. General Bernard Montgomery, in turn, replaced him.

Montgomery, who was even slower to attack than his predecessors, was appointed by Churchill, who acted on the advice of the commander, Imperial General Staff, Sir Alan Brooke.[8] Montgomery was a skilled practitioner of positional warfare. More so than his predecessors, excepting O'Connor, he seems to have had a good grasp of mechanized warfare. Montgomery preferred using overwhelming force rather than ambitious maneuvers, because it played to British strengths, and in the long run would spare the lives of his troops. Much to his credit, he kept his nerve as Rommel threatened Alexandria, and he managed eventually to draw the Afrika Korps into a battle of attrition that they could not possibly win.

Even so, it is by no means certain that Montgomery's predecessors would not have achieved the same result perhaps sooner and at less cost. That the convulsions at the top of the military command were unwise (however politically expedient in Churchill's mind) is thoroughly discussed in Correlli Barnett's book, *The Desert Generals*, a point of view endorsed by Rommel in 1944. "Viewed as a whole," Rommel recalls in his private papers, "it was a great mistake for the British to be continually replacing their Commander-in-Chief, and thus forcing the new man to learn the same bitter lessons all over again."

> The British commanders were capable soldiers; it was merely that some of them had pre-conceived ideas . . . which they would certainly have discarded after their first reverses. But they were always relieved of their command before they had the chance.[9]

To British troops in the line the effect of these changes must have been demoralizing. Over the course of the campaign many of Rommel's subordinate commanders were killed, severely wounded, or captured in battle (Generals Bismarck, von Thoma, Nehring, Cruewell, von Prittwitz, Stumme), making them heroes. Their British counterparts were cashiered, thus suggesting to British troops that Rommel was invincible and that their own commanders were incompetent.

Rommel's gifts, however, were counterbalanced by traits that undermined the strategic objectives he was under an obligation to realize. His astonishing success—which outweighed the disobedience that made them possible—affected British strategic decisions all out of proportion to the force he employed, but his achievements came to bedazzle his own senior commanders in ultimately harmful ways. Despite orders to fight a strictly defensive campaign, along with his subordination to the

Italian chief of staff—a sensible arrangement given that most of the Axis troops in theater were Italian—Rommel was seduced by other possibilities more in line with his temperament. Long before the German drive to the Caucasus was set down as a plan in the spring of 1942, Rommel wanted to capture the Suez Canal. Beyond that, a glance at the map put before him the prospect of driving to the Caspian Sea on a northeasterly axis. The supply problem could be mitigated, if not eliminated, by the capture of huge stocks in Alexandria and Cairo. This idea was not unwelcome to Hitler, he and Rommel discussed the idea in March 1942, as Hitler was formulating Case Blue, but the General Staff recognized Plan Orient for what it was: fantasy undiluted by the impositions of geography and logistics. Even so, the idea became increasingly attractive during the summer of 1942 as Rommel, working one operational miracle after another, reached the Egyptian border, and General von Kleist's First Panzer Army drove southward from Rostov, hindered more by shortages of fuel than by enemy action.

Rommel's conceit was fed not only by his own success against a numerically superior enemy who wilted before him, but also by the likely possibility of a decisive victory in southern Russia, and by the emerging support of the General Staff (no doubt goaded by Hitler), which would yield enthusiasm for a more aggressive strategy against the British in the Middle East. In August 1942, on the eve of the battle of Alam Halfa, where Rommel was defeated, General Warlimont, the deputy chief of staff, visited Rommel and "had spoken optimistically of 'Plan Orient.'" As late as December 1942, Hitler told General von Manstein that Stalingrad needed to be relieved rather than evacuated as a necessary step toward realizing Plan Orient. By that time, not only was Rommel forced to go on the strategic defensive for good, but there was no longer any justification for maintaining a large force in Africa indefinitely. The only possible value in keeping Afrika Korps in place was the same as that which called it into being: to tie down large numbers of allied troops as part of a skillfully executed withdrawal to Tunisia and, ultimately, to Italy. But Hitler's love of Wagnerian events kept him from seeing an evacuation as anything other than humiliation, and so he diverted troops and equipment that might have been used on the Eastern Front to support a lost cause. Far better (Hitler no doubt thought to himself) for German forces in Africa to go down fighting, even if doing so conformed to the strategic objectives of the Allies. Rommel's reputation thus provided Hitler with another justification—not that one was needed—to pursue a strategy that led to strategic disaster, not only in the Mediterranean, but in Russia as well.[10]

In other ways Rommel's genius frustrated his ambitions. Time always worked against him in the desert: if he stood still his enemy's strength would accumulate, while his own—dependent as it was on an uncertain supply line—would diminish. Most of the time Rommel solved the logistical problems that bedeviled his grand schemes by capturing British stocks of water, food, fuel, and vehicles. For him, offensive action was akin to the need of a predator to secure food, as much as it was part of a carefully reasoned strategic plan.

After a series of offensives against the El Alamein position ended in stalemate in July 1942, Rommel lunged for a final time at the Suez Canal by attacking British defensive positions at Alam Halfa in late August 1942. The battle was a crushing failure for the Afrika Korps. Assuming that the newly appointed head of Eighth Army, General Bernard Montgomery (who, apart from leading a division in France, had no reputation as a field commander) would act as other British commanders had previously, Rommel applied pressure at various points in the hopes of drawing Montgomery's mobile reserves into the open, where his panzers could defeat them in detail.

In the past Rommel had always managed to defeat forces that were much stronger than his own. His aggressiveness threw his opponents off balance and often prompted them to act impetuously. Commanding from the front, he was possessed of a remarkable ability to discern where his opponent was likely to give way, and direct his forces accordingly. But at Alam Halfa Montgomery's operational senses did not desert him. In Montgomery, Rommel confronted an opponent whose moves were always designed to limit his opponent's options, while increasing his own.

By drawing up defensive positions between El Alamein and the Quattara Depression thirty miles to the south, Montgomery closed off the option—Rommel's favorite—of attacking round the flank of his opponent while applying pressure elsewhere. Of greater importance, Montgomery did not surrender his judgment to momentary good news. Every decision was carefully thought through; Montgomery did not act until he was sure that a given engagement played to his material strengths and minimized Rommel's skill at feint and maneuver. This manner of proceeding, of course, illuminates the difference between a commander who remains self-possessed, and one overly eager for the big victory, the quick score. These opposing frames of mind, which can and often do exist at different times in the same person, will compel a commander to act differently on identical information. Montgomery was not overawed by Rommel's reputation, and this is perhaps at the root of his triumph.

Montgomery has been criticized (rightly, in most cases) as being too concerned with his public image. But at the Battle of Alam Halfa (August/September 1942) and El Alamein (October/November 1942), his appetite for glory was disciplined by his considerable intellectual ability and moral steadfastness. Montgomery recognized that his enormous advantage in men and material would be of no use—apart from gilding Rommel's reputation—unless he turned his opponent's strengths against him. In this battle, Rommel for once was a victim of his previous victories. His forces were too weak, he had neither strategic reserves nor logistical resources to fight a pitched battle, and many of his troops had endured privations imposed on them by desert warfare for about a year and a half to the point where their physical endurance, if not their morale, was approaching its limit. Much to Rommel's consternation, the battle of Alam Halfa took on the character of a chess game, where victory cleaves to the more discerning and patient combatant, rather than a scratch brawl in which a quick punch in the right spot allows the weaker belligerent to demoralize his opponent and so prevail.

It would be a mistake to characterize the battles near El Alamein between August and November 1942 as examples of where a general skilled at maneuver was overwhelmed by brute force. There is some truth here, but at heart Rommel was out-thought by Montgomery. Rommel accepted battle at a place and under circumstances of Montgomery's choosing. Despite immense pressure, aggravated by the psychological weight of defeats accumulated over the previous eighteen months, Montgomery's troops stuck to their guns, which was hardly surprising, given that the first priority for Montgomery when he took command of Eighth Army in August 1942 was to rehabilitate the morale of the rank and file.

Bernard Montgomery is easy to dislike. In so many ways he embodies traits that would end his career nowadays at the rank of lieutenant, irrespective of whatever military aptitude he might demonstrate. Taking stock of his character and conduct proves that he would be out of sorts when it comes to our standard-issue command culture. Vain, priggish, bloodlessly irritable, abstemious, aloof, and not inclined to give due credit to allies or subordinates, Monty was allergic to what we consider the essence of being a "team player." His prickly nature was not lost on his contemporaries; Winston Churchill, in referring to his social rather than physical stature, is said to have called him "a little man on the make." Nevertheless, in North Africa Montgomery proved to be, if not an avatar of military leadership, well prepared by training and temperament to confront Rommel. Montgomery took charge of Eighth Army, but not with the idea of mimicking Rommel's method for he knew his troops were not the equals of the Afrika Korps in tactical matters. Rather, he began by rehabilitating the army's morale. He correctly discerned that much of Rommel's mythic strength was merely a reflection of the receding self-confidence of his opponent. British troops were not wretchedly or truculently demoralized—as were the Allied armies in France in 1917—but dispirited and apprehensive. Their stalwart courage had up until then yielded nothing other than the respect Rommel paid them, which is every brave soldier's due even in defeat.[11]

The first thing Montgomery did was deviate from the standard-issue uniform; his gaily decorated hats and walking shorts that displayed his "white knees" immediately caught the eye of his soldiers. It was a conspicuous gesture meant to communicate to them that "things will from now on be different." Monty spent an enormous amount of time visiting his troops—each morning and afternoon were devoted to this. His speeches to the rank and file were free of ornament; he neither flattered nor hectored his soldiers.[12] The duty of the Eighth Army, he told them, was to kill Germans or die trying. On his watch there would be no withdrawal. Montgomery ordered destroyed, and refused to countenance further, any plan incorporating a retreat to Alexandria. He also elevated the status of the chaplains, for the role of transcendent faith was, in his mind, an essential component of a battle-ready force. In return, his troops gave him a degree of loyalty that none of his predecessors enjoyed. His many faults notwithstanding, Montgomery reconstituted the spirit of his army and that helped them more than any material factor to defeat a resourceful, determined, and courageous enemy.

Nowadays, much of a commander's day is absorbed in managerial minutiae, which springs from our conflation of the corporate executive and the commander—a habit of mind that does an injustice to both occupations. Commanders devote an immense amount of time moving information up and down the chain of command, responding to higher headquarters tasking, and meeting with one's increasingly corpulent staff organizations. Montgomery, by contrast, left the administrative details of command to his small staff; his success puts the lie to the idea that command requires a meticulous grasp of enormous stores of information.

Montgomery (the same holds true for Rommel) did not need to consult books with such titles as "The Top Ten Habits of the World's Top Ten Administrative Wizards," or "How to Accomplish in Twelve Seconds What Normally Takes Two Minutes," and the like. Both generals had a conceptual understanding of the sort of campaign they wanted to fight, informed by a deep and respectful understanding of their profession, which was fortified by a singleminded devotion to achieving their operational aims. "Thoughtful" and "efficient" were for them not antagonistic terms. Rommel frequently consulted a German translation of General Wavell's essays on strategy. Montgomery knew that victory hinged on his ability to out-think Rommel. Both generals would find our system-of-systems approach a baffling thing.

For all of their huge ambition, both men were anything but self-centered, which is often an apt synonym for keen career-mindedness. Rommel and Montgomery could be, and often were, brutally frank. But this is a refreshing trait when compared with the contrived congeniality (fierce egotism masquerading as good character) that our culture misconstrues as a necessary component of leadership. There was nothing frigidly ceremonial about the interaction of Rommel and Montgomery with their troops. They relished being among them, and their welfare was always forefront in their minds. The authority of both Rommel and Montgomery rested not on the decisions of promotion boards scrutinizing jargon-laden, stereotyped performance reports, but on the character of the man wielding it.[13]

The generalship of Montgomery and Rommel is a fit subject for emulation, if not hagiographical imitation. But a proper appreciation of the interaction of circumstance and character exemplified in the North African campaign takes a vigorous and disciplined species of imagination, something that our current efforts at transformation, and our approach to educating, training, and promoting officers, holds at a sharp discount. It is not that reforms based on what is discussed in this book will produce a latter-day Rommel or Montgomery, though that is much more likely than if we continue the current status quo. Rather, our exaltation of the technician and the manager will keep us from understanding, let alone profiting from, the strengths as well as frailties of the Desert Fox and Monty—a habit of mind that we dare not concede to a prospective enemy.

Concluding Thoughts

Our technology-besotted efforts to reform the U.S. military may turn out to frustrate what is at heart a praiseworthy and indeed urgent task. It is even possible that we are setting ourselves up for disaster in spite of our commendable if misbegotten zeal to make our combat power more effective than it already is. What goes under the name of transformation today is a counterfeit of the real thing. It seems that we've taken to heart the canard that military leaders fail because they prepare for the next war by planning to refight the last one. But is there any other reliable guide to the future other than studying how things have gone in the past?

The antidote to our manic devotion to futurism is to ponder what history sets before us, which among other things proves that innovation cannot be manufactured by a bureaucracy, nor can it be summoned into being by committee. History shows that wars are won or lost not because one side produced a bureaucratic arrangement or piece of technology that, in the abstract, was superior to that of an opponent. Weapons and masses of troops do eventually tell, as the victory of the Red Army over the *Wehrmacht* amply demonstrates. But this outcome could not, ultimately, have been reversed by a shift in material factors alone. The Soviet Union was transformed into one of the two great powers and went on to menace the civilized world for five decades, because Germany's strategic calculations were based on faulty ideas, not to mention iniquitous habits of mind that could not help but unite much of the world—if only momentarily—against her. France was overrun not because her military equipment was inadequate (some of her weaponry was quite good), but because her military leadership, reflecting the malaise of the broader culture, was defective in moral strength and hostile to ideas that clashed with received opinion.

It is not enough for us to congratulate ourselves for having avoided many of the mistakes that have afflicted other armies across time. Rather we should look to history—a record of how things go—as a means of determining how best to reform the military. Instead of embracing, in the manner of a desperate lover, slogans and gadgetry and bureaucratic enchantments, we ought to consider more substantive, if less glamorous, ways of rehabilitating our armed forces. A first step in the right direction would be to at least consider the possibility that success in war pivots not on inert objects or systems, but on the brain of an army. The wellsprings of transformation can be found in the moral and intellectual constitution of commanders and soldiers. And the best way to cultivate officers who are capable of facing whatever belligerent future happens to be percolating just beyond the horizon is to recast our educational and promotion policies so that they become as estimable as the valor and tactical skill of our troops. If we fail to give proper attention to fortifying the judgment of those who actually do the fighting, then all of the bedazzling gadgetry that science can deliver to us, and all the clever slogans and bureaucratic wizardry that the connoisseurs of Powerpoint™ slideshows find so bewitching, will add up to zero. Or, what is more likely, they will end up as trophies that will burnish the reputation of twenty-first-century successors to Rommel and von Manstein.

Appendix

JOINT ADVANCED WARFIGHTING SCHOOL
NATIONAL SECURITY STRATEGY AND NATIONAL SECURITY SURPRISE

WAR FROM TECTONIC DRIFT—a Case Study

TH6116A
SYLLABUS (Academic year 2005–2006)

FACULTY GUIDANCE

1. PURPOSE AND SCOPE

  a. This block, TH6116, looks at the reality of the outbreak of war, sometimes anticipated, sometime not. It begins with the case study of a drift to war, now historically perceptible, the United States and Japan [sic]. It follows with two lessons that look at the background of more recent, still unfolding, situations—confrontation with the Islamic world and the coming of age of China. It is a foundation and lead-in to the preliminary lessons of the next field of study—the contemporary process of the formulation of national security strategy.

  b. This first lesson is a case study of the long path to conflict between the United States and Japan in the Pacific. It is the tale of a drift to war, of evolving national security interests and derivative military and theater strategies. The study looks at the fundamental interests of the two nations and the convergent causes that ultimately resulted in Japan's surprise attack at Pearl Harbor in 1941. For all but the very end of the journey there was not a will to war, yet the flow of tectonic-like circumstance brought it on. Two hours are given to the case study. The third one is a historical inventory of other perceptible categories for the onset of war, e.g. essentially complete strategic surprises, well-calculated initiatives, chronic underestimations.

2. LESSON OBJECTIVES

a. Analyze for cause and effect the evolution of the national security strategies of the United States and Japan that brought the two countries to hostilities in December 1941.

b. Analyze for cause and effect the evolution of the military strategies of each nation up to the outbreak of war and then to Midway.

c. Analyze for cause and effect the campaign strategies of both countries, Pearl Harbor to Midway.

d. Evaluate for other, past, categorizable paths to war.

3. LESSON DEVELOPER

XXXXXXX

4. STUDENT PARTICIPATION REQUIREMENTS

Complete readings; participate in the seminar's discussion.

READINGS

a. James, D. Clayton. "American and Japanese Strategies in the Pacific War" in *Makers of Modern Strategy from Machiavelli to the Nuclear Age.* Edited by Peter Paret. Princeton: Princeton University Press, 1986. (Selected Readings)
—— pp. 703–713. "American and Japanese Strategies in the Pacific War" (part)

b. Morton, Louis. "Germany First: The Basic Concept of Allied Strategy in World War II" in *Command Decisions.* Edited by Kent Roberts Greenfield. Washington, D.C.: U.S. Government Printing Office, 1960. (Selected Readings)
—— pp. 11–38. "Germany First" (part)

c. Morton, Louis. "Japan's Decision for War" in *Command Decisions.* Edited by Kent Roberts Greenfield. Washington, D.C.: U.S. Government Printing Office, 1960. (Selected Readings)
—— pp. 99–124. "Japan's Decision for War"

d. Hart, B. H. Liddell. *History of the Second World War.* NY: G. P. Putnam's Sons, 1970. (Selected Readings)
—— pp. 343–353. "The Tide Turns in the Pacific" (part)

5. ADDITIONAL STUDENT REQUIREMENTS

None.

6. TEACHING PLAN

Case study, lecture, guided discussion

7. STUDENT EVALUATION/GRADING

The immediate criterion is the quantity of quality contribution to the seminar's discussion in class. Further, the student's ability to address this lesson in synthesis is subject to inclusion in a comprehensive examination.

8. QUESTIONS FOR DISCUSSION

Refer back to lesson objectives.

9. OPMEP LEARNING AREAS ADDRESSED

Area 1—National Security Strategy, Systems, Processes and Capabilities; Area 2—Defense Strategy, Military Strategy, and the Joint Operations Concepts; Area 3—Theater Strategy and Campaigning with Joint, Multinational and Interagency Assets; Area 6—Joint Strategic Leader Development.

Notes

Chapter 1

1. W. B. Allen, ed. *George Washington: A Collection* (Indianapolis: Liberty Fund, 1988), 40.
2. Alexis de Tocqueville, *Democracy in America,* Harvey Mansfield & Delba Winthrop, trans. & ed. (Chicago: Univ. of Chicago P., 2000), 390; James Bryce, *The American Commonwealth* (Indianapolis: Liberty Fund, 1995), 1193–1194.
3. J. D. Hittle, *The Military Staff* (Harrisburg, PA: Military Service Publishing, 1949), 150–151.
4. Steven T. Ross, *American War Plans 1890–1939* (London: Frank Cass, 2002), 7*ff.*
5. Qtd. in Hittle, *Military Staffs,* 178. Also see Peter R. DeMontravel, *A Hero to His Fighting Men: Nelson A. Miles, 1839–1925* (Kent, OH: Kent State Univ. P., 1998), 307–309; 344*ff.*
6. Carl von Clausewitz, *On War,* Peter Paret, ed. & trans. (Princeton: Princeton Univ. P., 1976), 100–112; Christopher Bassford, *Clausewitz in English* (New York: Oxford Univ. P., 1994), 50–55.
7. Philip C. Jessup, *Elihu Root* (New York: Dodd, Mead & Co., 1938), *passim.*
8. Paul M. Kennedy, ed. *The War Plans of the Great Powers, 1880–1914* (London: George Allen & Unwin, 1979), 23*ff.* Also see Barbara Tuchman, *The Zimmermann Telegram* (London: Folio Society, 2004).
9. For an excellent contemporary criticism of the age, see Jose Ortega y Gasset, *The Revolt of the Masses* (New York: W. W. Norton, 1932).
10. Henry Gole, *Road to Rainbow* (Annapolis: Naval Institute Press, 2003), *passim;* Max Hastings, *Armageddon* (New York: Knopf, 2004), 8. In addition

to the works of Stephen Ambrose, an excellent book that focuses on the valor of America's citizen soldiers is: Alex Kershaw, *The Longest Winter: The Battle of the Bulge and the Epic Story of WWII's Most Decorated Platoon* (Cambridge, MA: Da Capo, 2004).

11. *Capstone Concept for Joint Operations,* Aug 05, C-2.
12. *JP 5-0,* April 2006, III-13.
13. George S. Patton, *War as I Knew It* (Boston: Houghton Mifflin, 1947), 116.
14. Basil Liddell Hart, *The Second World War* (London: Cassell, 1970), 42–48; Winston Churchill, *The Second World War* (Boston: Houghton Mifflin, 1983), I: 538*ff*; Richard Condon, *The Winter War* (New York: Ballantine 1972) *passim*; William Trotter, *Frozen Hell* (Chapel Hill: Algonquin Books, 1991), *passim*; *The World at War*, Jeremy Isaacs, prod. (A&E Home Video, 2004), vol. II; Seaton, *Stalin As Military Commander* (New York: Praeger, 1976), *passim.* Also see Alan Bullock, *Hitler & Stalin: Parallel Lives* (New York: Knopf, 1991), 659*ff.*, and *passim.*
15. Daniel Hughes, ed., *Moltke on the Art of War.* (Novato, CA: Presidio, 1993), 36.
16. Donald Rumsfeld, "Beyond Nation Building," 14 Feb 2003; Gordon & Trainor, *Cobra II* (New York: Pantheon, 2006) 152.
17. *Cobra II,* 26–27 and *passim.*
18. Ricks, *Washington Post,* 25 December 04; also see Thomas Mackubin Owens, "Rumsfeld & his Critics," *Weekly Standard,* 3 May 06.
19. *Cobra II,* 486.
20. *Cobra II,* 128, 150, 314.
21. Tommy Franks, *American Soldier* (New York: ReganBooks, 2004), 525, 545, 550 and *passim.*
22. Michael E. O'Hanlon, "Iraq Without a Plan," *Policy Review,* Dec 04/Jan 05.
23. *Cobra II,* 145–147.
24. Conrad C. Crane & W. Andrew Terrell, *Reconstructing Iraq* (Carlisle, PA: SSI, 2003), iv, 54.
25. Earl F. Ziemke, *The U.S. Army in the Occupation of Germany 1944–1946* (Washington, D.C.: U.S. Army Center for Military History, 1990), 106*ff.*
26. *Foxnews.com,* 24 & 30 March 2006.
27. Barbara Tuchman, *The March of Folly* (New York: Knopf, 1984), 5.
28. *On War,* 194–195.
29. Ian Buruma, *Inventing Japan: 1853–1964* (New York: Modern Library, 2003), 132 and *passim.*
30. *The Zimmermann Telegram,* 158–159; 170–171.
31. Ziemke, *Occupation,* 320. National Security Council, *The National Strategy for Victory in Iraq,* November 2005. Available at http://www.whitehouse.gov/infocus/iraq/iraq_national_strategy_20051130.pdf.
32. Foxnews.com, "FBI Puts Local Officials on Notice About Extremists Trying to Sign Up to be School Bus Drivers," 16 March 2007. Accessed 19 March 2007. Available at http://www.foxnews.com/story/0,2933,259168,00.html.
33. Nicholas E. Reynolds, *Basrah, Baghdad & Beyond* (Annapolis: Naval Institute Press, 2005), 109.

34. James Monroe, "The Monroe Doctrine," rptd. in *The Annals of America* (Chicago: Encyclopedia Britannica, 1976), V: 73–76.
35. John Lewis Gaddis, *Surprise, Security, & the American Experience* (Cambridge: Harvard Univ. P., 2004), 64*ff*.
36. Tony Zinni, *The Battle for Peace* (New York: Palgrave Macmillan, 2006), 143.
37. *National Security Strategy of the United States*, March 2006 (*NSS*), 1–3.
38. *NSS*, 33.
39. *NSS*, 45.
40. Michael Smerconish, "Listen to Lehman," *National Review Online*, 15 April 2004; Benjamin Netanyahu, *Fighting Terrorism* (New York: Farrar, 2001), *passim*.
41. Stanley Kurtz, "Polygamy vs. Democracy," *Weekly Standard*, 5 June 2006, 20 and *passim*.
42. Barbara Tuchman, *The Proud Tower* (London: Folio Society, 1995), 433–434.
43. Walter Bagehot, *The Historical Essays*, Norman St. John-Stevas, ed. (London: *The Economist*, 1968), IV: 48–49; also see Raymond Postgate, *1848: The Story of a Year* (New York: Oxford Univ. P., 1956).
44. Samuel P. Huntington, *Who Are We?* (New York: Simon & Schuster, 2004), 209*ff*. and *passim*); George Will, "On Bilingual Ballots," *Townhall.com*, 25 May 06.
45. Tony Allen-Mills, "Pizza Pope Builds a Catholic Heaven," *The Sunday Times*, 26 Feb 06. Available at http://www.timesonline.co.uk/article/0,,2089-2058771,00.html.
46. Jimmy Carter, *L.A. Times* op-ed, 14 Nov 2005; speeches by Albert Gore, and William Clinton, 7th Jeddah Economic Forum, 11–13 Feb 2006.
47. Kenneth Anderson, "Foreign Law & the U.S. Constitution," *Policy Review*, June/July 2005.
48. "Google to censor itself in China," *CNN.Com*, 25 Jan 06.
49. For a discussion of the rise of Islamic militancy in Europe see Mark Steyn's *America Alone* (2006) and Bruce Bawer's *While Europe Slept* (2007).
50. Sam Harris, "Head-in-the-Sand Liberals: Western civilization really is at risk from Muslim extremists," *Los Angeles Times*, 18 Sep 2006. Available at http://www.samharris.org/site/full_text/the-end-of-liberalism/.
51. Joseph Loconte, "The World of Red Ken: The Mayor of London Debates Daniel Pipes," *Weekly Standard*, 5 Feb 2007.
52. George Walden, *Time to Emigrate?* (London: Gibson Square, 2006); also see Arthur Herman, "The Strange Death of the Royal Navy," *New York Post*, 14 Jan 07.
53. *Diplomacy* (New York: Simon & Schuster, 1994), 811.
54. *Germany & the Next War*, Allen H. Powles, trans. (New York: Longmans, Green, & Co., 1914).
55. See Basil Liddell Hart, *Strategy* (New York: Praeger, 1954), 23–26.
56. *QDR*, 42, 64.
57. The drawing, "En Alsace!," originally appeared in *L'Illustration*, 15 August 1914. See Barbara Tuchman, *The Guns of August* (London: Folio Society, 1990), illustrations between 164–165.

58. Tuchman, 30*ff.*
59. See Basil Liddell Hart, *Strategy* (New York: Praeger, 1954).
60. *QDR*, 4.
61. *QDR*, 38; also see 20.
62. *QDR*, 84.

Chapter 2

1. A portion of this chapter first appeared as an essay, "Transformation Ballyhoo," published in *USNI: Proceedings,* September 2006.
2. Office of Force Transformation, *Military Transformation: A Strategic Approach,* 2.
3. *Military Transformation,* 6.
4. *Military Transformation,* 8.
5. *Military Transformation,* 9.
6. Douglas Macgregor, *Transformation Under Fire* (Westport, CT: Praeger, 2003), 18.
7. *Transformation Under Fire,* 24–25.
8. For details on British and American armored fighting vehicles see Chamberlain & Ellis, *British & American Tanks of World War Two* (London: Arms & Armour, 2002); Kenneth Macksey, *Tank Force: Allied Armor in World War II* (New York: Ballantine, 1970). For Soviet tank specifications see John Milsom, *Russian Tanks 1900–1970: The Complete Illustrated History of Soviet Armoured Theory & Design* (Harrisburg, PA: Stackpole Books, 1971); Douglas Orgill, *T-34: Russian Armor* (New York: Ballantine, 1974); also see http://www.wwiivehicles.com. For details on German armored fighting vehicles see Chamberlain and Doyle, *Encyclopedia of German Tanks of World War Two* (London: Arms & Armour Press, 2001); McCarthy, Peter, & Mike Syron, *Panzerkrieg: The Rise & Fall of Hitler's Tank Divisions* (New York: Carroll & Graf, 2002); Douglas Orgill, *German Armor* (New York: Ballantine, 1974); Bryan Perrett, *Knights of the Black Cross: Hitler's Panzerwaffe & its Leaders* (New York: St. Martin's, 1986). Matthew Cooper & James Lucas, *Panzer: The Armoured Force of the Third Reich* (New York: St. Martin's, 1978).
9. See John Weeks, *Infantry Weapons* (New York: Ballantine, 1971); Ian V. Hogg, *Barrage: The Guns in Action* (New York: Ballantine, 1970), 80.
10. See Albert and Joan Seaton, *The Soviet Army 1918 to the Present* (New York: New American Library, 1986); Albert Seaton, *Stalin As Military Commander.*
11. See Douglas Orgill, *German Armor* (New York: Ballantine, 1974).
12. Ian Hogg, *The Guns: 1939–1945* (New York: Ballantine, 1970), 63–64; John Weeks, *Men Against Tanks* (New York: Mason/Charter, 1975), 51*ff.*
13. Weeks, *Infantry Weapons,* 106–108.
14. For a cogent discussion of this subject see Mark Brilakis, "Martian Alert: Who's the Enemy," *USNI: Proceedings,* Jan 06, 27–40.
15. *Transformation Under Fire,* 22.

16. See Basil Liddell Hart, *Strategy: The Indirect Approach* (New York: Praeger, 1954).

17. Hastings 19–22; Brian Horrocks, *A Full Life* (London: Collins, 1960), 203–206; Geoffrey Perret, *Eisenhower* (New York: Random House, 1999), 314–320.

18. *Transformation Under Fire,* 200.

19. Elizabeth Kier, "Culture & Military Doctrine: France Between the Wars," *International Security,* Spring 1995, 65–93.

20. Alistair Horne, *To Lose a Battle: France 1940* (Boston: Little, Brown & Co., 1969), 22*ff.*

Chapter 3

1. A portion of this chapter originally appeared as an essay, "Time for a First-Rate Joint Staff College," published in *USNI: Proceedings,* December 2006.

2. Spenser Wilkinson, *The Brain of an Army* (London: Constable & Co., 1913), 5; Bassford, *Clausewitz in English,* 83*ff.*

3. Wilkinson, 153–154.

4. Martin van Creveld, *The Training of Officers* (New York: Free Press, 1990), 69*ff.*

5. Basil Liddell Hart, *The Remaking of Modern Armies* (Boston: Little, Brown & Co., 1928), 177.

6. Mackubin Thomas Owens, "Rumsfeld & His Critics," *The Weekly Standard,* 3 May 2006. A particularly thoughtful debate on the subject can be found in the June 2006 issue of *USNI: Proceedings.*

7. Colin Gray, *Another Bloody Century* (London: Weidenfeld & Nicolson, 2005), 189, 200–201.

8. See Appendix A.

9. Geoffrey Perret, *Eisenhower* (New York: Random House, 1999), 80–81.

10. Delta State University official Web site. Available at http://www.deltastate.edu/pages/1.asp.

11. "Blind Science," *National Review,* 4 July 05.

12. Bagehot, *Historical Essays,* III: 2.

Chapter 4

1. Hart, *Why Don't We Learn From History?* (New York: Hawthorn Books, 1971), 19.

2. Hart, *Why Don't We Learn From History?* 17.

3. Horne, *To Lose A Battle,* 581–583.

4. For competing views on the effectiveness of the resistance, see Max Hastings, *Das Reich: Resistance & the March of the 2nd SS Panzer Division Through France, June 1944* (New York: Henry Holt, 1982), and John Keegan, *The Second World War* (New York: Viking, 1990), 483–496. For a discussion of Anglo-American relations at this time, see Martin Gilbert, *Churchill: A Life*

(New York: Henry Holt, 1991), 687ff; Samuel Eliot Morrison, *The Oxford History of the American People* (New York: Oxford Univ. P., 1965), 996ff. Also see Paul Johnson, *A History of the American People* (New York: HarperCollins, 1997), 775ff.

    5. William Shirer, *The Rise & Fall of the Third Reich* (New York: Simon & Schuster, 1960), 894–895; Alexis de Tocqueville, *Democracy in America*, Harvey C. Mansfield, ed. & trans. (Chicago, Univ. of Chicago P., 2000), 573–576.

    6. Hart, *Second World War*, 81–86; J. F. C. Fuller, *Decisive Battles of the Western World* (London: Cassell, 2001), III: 527; for the effect of the Dieppe raid see Churchill, *Second World War*, IV: 509–511.

    7. Horne, *To Lose a Battle*, 56; John Keegan, *The First World War* (New York: Knopf, 1999), 423.

    8. John H. Reilly, *Jean Giraudoux* (Boston: Twayne Publishers, 1978), 145; also see Laurent LeSage, *Jean Giraudoux: His Life & Works* (State College: Penn. State Univ. P., 1959), 47ff.

    9. *Tiger at the Gates*, Christopher Fry, trans. (New York: Oxford Univ. P., 1955), 67. This edition adapted a line from the play as its title.

    10. *Under Fire*, Robin Buss, trans. (New York: Penguin, 2004), 315–316.

    11. Horne, *To Lose a Battle*, 55–56.

    12. See Erich Maria Remarque, *All Quiet on the Western Front*, A. W. Wheen, trans. (New York: Ballantine, 1991); Christine R. Barker & R. W. Last, *Erich Maria Remarque* (London: Oswald Wolff, 1979), 17ff.

    13. *Storm of Steel* (New York: Fertig, 1996), 316–317.

    14. Horne, *To Lose a Battle*, 11ff.

    15. Horne, *To Lose a Battle*, 38.

    16. Albert Seaton, *The German Army 1933–1945* (New York: St. Martin's, 1982), 71ff; Horne, *To Lose a Battle*, 39–41; Kenneth Macksey, *From Triumph to Disaster* (Mechanicsburg, PA: Stackpole Books, 1996), 64; Alfred Price, *The Luftwaffe* (New York: Ballantine, 1969), 11–14; Mitcham, *Panzer Legions*, 37.

    17. Michael Howard, *The Franco-Prussian War* (Norwalk, CT: Easton Press, 1994), 8–57.

    18. Tuchman, *The Proud Tower*, 157ff; Rudyard Kipling, *Barrack Room Ballads* (New York: Alex Grosset & Co., 1899), 101.

    19. Keegan, *First World War*, 330–331.

    20. Charles De Gaulle, *The Army of the Future* (New York: J. B. Lippincott, 1941), 37.

    21. Horne, *To Lose a Battle*, 81–84.

    22. *The Army of the Future*, 47.

    23. *The Army of the Future*, 84–85.

    24. *The Army of the Future*, 78–79, 104.

    25. *The Army of the Future*, 170–171.

    26. William Shirer, *The Collapse of the Third Republic* (New York: Simon & Schuster, 1969), 186, 397–398, 761.

    27. Michael Howard, ed., *The Theory & Practice of War* (London: Cassell, 1965), 135ff.

    28. Kenneth Macksey, *Tank Force* (New York: Ballantine, 1970), 24ff.

    29. Hart, *Remaking of Armies*, 248–249.

30. Kenneth Macksey, *From Triumph to Disaster*, 55–60; Richard Humble, *Hitler's High Seas Fleet* (New York: Ballantine, 1971), 25–29; also see Barry Leach, *The German General Staff* (New York: Ballantine, 1973), 8*ff*.

31. Kenneth Macksey, *Guderian: Panzer General* (London: Greenhill Books, 2003), 4–6.

32. See Macksey, *Guderian*, chapters two–six; Peter McCarthy & Mike Syron, *Panzerkrieg* (New York: Carroll & Graf, 2002), 19*ff*; Heinz Guderian, *Achtung! Panzer!*, Christopher Duffy, trans. (London: Arms & Armour Press, 1992), *passim*.

33. *Guderian*, 44, 69.

34. The original Case Yellow failed to consider "the scope for maneuver open to a bold and resolute enemy commander," writes General von Manstein in his memoirs. "One had no right to assume that such leadership would be lacking, particularly in view that General Gamelin [the French Army Commander-in-Chief] enjoyed with us. He certainly made an excellent impression on General Beck when the latter visited him before the war." See Erich von Manstein, *Lost Victories* (Chicago: Regnery, 1956), 102.

35. Hugh Trevor-Roper, *Blitzkrieg to Defeat: Hitler's War Directives* (New York: Holt, Rinehart & Winston, 1964), 21.

36. "War Directive 6" was issued on 9 October 1939; The French General Staff began war planning against the Germans began on 26 September (*Blitzkrieg to Defeat*, 13; Theodore Draper, *The Six Weeks War*, 24).

37. For a full discussion of these subjects see Alistair Horne, *To Lose a Battle*, *passim*.

38. See Horne, *To Lose a Battle*, 130–131.

39. Horne, *To Lose a Battle*, 493–509; *Panzerkrieg* 85–86; Kenneth Macksey, *The Shadow of Vimy Ridge* (Toronto: Ryerson, 1965), 204*ff*.

40. Fuller, *Decisive Battles* III: 538–589.

Chapter 5

1. Hastings 8; Paul Adair, *Hitler's Greatest Defeat* (London: Arms & Armour Press, 1994), *passim*; Alex Buchner, *Ostfront 1944* (West Chester, PA: Schiffer Publishing, 1991), 141–218; 239–304. Also see Rolf Hinz, *To the Bitter End*, Frederick P. Steinhardt, trans. (Solihull: Helion & Co., 2005), *passim*. For a discussion of the fortress cities see Christopher Duffy, *Red Storm Over the Reich* (Edison, NJ: Castle Books, 2002) and Jurgen Thorwald, *Flight in the Winter: Russia Conquers, January to May 1945*, Fred Wieck, ed. & trans. (New York: Pantheon, 1951).

2. Alan Clark, *Barbarossa* (New York: William Morrow & Co., 1965), 287*ff*.

3. William Craig, *Enemy at the Gates* (Old Saybrook, CT: Konecky & Konecky, 1973), 191.

4. Craig, 190–191, 343.

5. See Robert J. Kershaw, *War Without Garlands* (New York: Sarpedon, 2000), 137–138.

6. Antony Beevor, *Stalingrad* (New York: Penguin, 1998), 357*ff*; Craig 317–319; Geoffrey Jukes, *Stalingrad* (New York: Ballantine, 1968), 149.

7. Craig, xiv–xv; Heinrich Gerlach, *The Forsaken Army* (London: Weidenfeld & Nicolson, 1958), 384.
8. Craig, 390*ff*; 435–437.
9. Bullock, *Hitler & Stalin*, 780–783.
10. Hart, *Second World War*, 243*ff*.
11. Hart, *Second World War*, 260–261.
12. An excellent first-hand account of the fighting in the Crimea is Gottlob Biedermann, *In Deadly Combat* (Lawrence, KS: University Press of Kansas, 2002).
13. David Glantz, *Colossus Reborn* (Lawrence, KS: Univ. Press of Kansas, 2005), 26*ff*.
14. Tieke, Wilhelm. *The Caucasus & the Oil* (Winnipeg: J. J. Fedorowicz, 1995), 359*ff*.
15. *Panzerkrieg*, 148–150; Clark, *Barbarossa* 259–264; Accessed 1 June 2006, available at http://www.11thpanzer.com. Balck's biography was written by Colonel David T. Zabecki, U.S. Army Reserve.
16. Erhard Raus, *Panzer Operations*, Steven Newton, ed. & trans. (Cambridge, MA: Da Capo, 2003), 137*ff*.
17. Raus, 137*ff*; Craig, 236–241; Jukes, 136–141; *Panzerkrieg*, 145*ff*.
18. Beevor, *Stalingrad*, 16; Manstein, 13–16, 361–62.
19. Joachim Wieder, *Stalingrad: Memories & Reassessments* (London: Arms & Armour Press, 1993), 159.
20. Walter Goerlitz, *Paulus & Stalingrad* (London: Meuthen, 1963), 11.
21. *Paulus*, 12.
22. *Paulus*, 17.
23. *Paulus*, 27.

Chapter 6

1. *World at War, vol. 3: Desert War*; Macksey, *Afrika Korps* (New York: Ballantine, 1968), 75, 120.
2. Richard Holmes, *Bir Hakim* (New York: Ballantine, 1971), 135–136; David Fraser, *Knight's Cross* (New York: HarperCollins, 1993), 307*ff*.
3. Wolfgang Schneider, *Tigers in Combat I* (Mechanicsburg, PA: Stackpole Books, 2004), 193; Peter Elstob, *Condor Legion* (New York: Ballantine, 1973), 122.
4. For a full account of the battle see Kenneth Macksey, *Beda Fomm* (New York: Ballantine, 1971).
5. Rick Atkinson, *An Army at Dawn* (New York: Knopf, 2004), 465–467.
6. Brian Horrocks, *A Full Life* (London: Collins, 1960), 132–133. Not all American antitank artillery was of domestic design. During the mid-1930s the first mass-produced antitank gun, the 3.7cm M3, was essentially a copy of the German PAK 35/36. As part of the Lend-Lease Agreement, the British provided the American Army with the drawings for their six-pounder, which was produced as the M1 5.7cm antitank gun (Hogg, *The Guns*, 58–60).
7. Churchill, *Second World War*, III: 64.
8. Alun Chalfont, *Montgomery of Alamein* (New York: Atheneum, 1976), 130–131.

9. Erwin Rommel, *The Rommel Papers,* B. L. Hart, ed. (New York: Harcourt, Brace and Co., 1953), 521.

10. Fraser, 363, 394, and *passim.*

11. Chalfont, 218.

12. See Montgomery's speech to Eighth Army officers, 13 August 1942 in *Lend Me Your Ears: Great Speeches in History,* William Safire, ed. (New York: Norton, 2004), 158–161.

13. Fraser, 275.

Works Consulted

Abbott, Wilbur C. *The Expansion of Europe: A History of the Foundations of the Modern World.* New York: Henry Holt, 1918.

Adair, Paul. *Hitler's Greatest Defeat: The Collapse of Army Group Centre, June 1944.* London: Arms & Armour Press, 1994.

Alger, John I. *The Quest For Victory: The History of the Principles of War.* Westport, CT: Greenwood P., 1982.

Allen, W. B., ed. *George Washington: A Collection.* Indianapolis: Liberty Fund, 1988.

Anderson, Kenneth. "Foreign Law & the U.S. Constitution." *Policy Review.* June/July 2005. Available at http://www.policyreview.org/jun05/anderson.html.

Atkinson, Rick. *An Army at Dawn: The War in North Africa, 1942–1943.* New York: Henry Holt, 2002.

Bagehot, Walter. *The Historical Essays.* Vols. 3 & 4 of the *Collected Works.* Norman St. John-Stevas, ed. London: *The Economist,* 1968.

Barbusse, Henri. *Under Fire.* Robin Buss, trans. New York: Penguin, 2004.

Barker, A. J. *German Infantry Weapons of World War II.* London: Arms & Armour Press, 1969.

Barker, Christine R. and R. W. Last. *Erich Maria Remarque.* London: Oswald Wolff, 1979.

Barnett, Correlli. *The Desert Generals.* Bloomington, IN: Indiana University Press, 1982.

Barzun, Jacques. *The Culture We Deserve.* Middletown, CT: Wesleyan University Press, 1989.

———. *From Dawn to Decadence: 500 years of Western Cultural Life.* New York: HarperCollins, 2000.

———. *A Jacques Barzun Reader.* Michael Murray, ed. New York: HarperCollins, 2002.
Bassford, Christopher. *Clausewitz in English: The Reception of Clausewitz in Britain & America, 1815–1945.* New York: Oxford University Press, 1994.
Beevor, Antony. *The Fall of Berlin 1945.* New York: Viking, 2002.
———. *Stalingrad: The Fateful Siege: 1942–1943.* New York: Penguin, 1998.
Boyne, Walter J. *Operation Iraqi Freedom: What Went Right, What Went Wrong, and Why.* New York: Tom Doherty, 2003.
Bryce, James. *The American Commonwealth.* 2 vols. Indianapolis: Liberty Fund, 1995.
Buchner, Alex. *Ostfront 1944: The German Defensive Battles on the Russian Front, 1944.* West Chester, PA: Schiffer Publishing, 1991.
Buckley, William F. "It Didn't Work." *National Review Online.* 24 February 2006. Available at http://www.nationalreview.com.
Bullock, Alan. *Hitler & Stalin: Parallel Lives.* New York: Knopf, 1992.
Bunting, Josiah. "Generation X and the Study of History." *Forging the Sword: Selecting, Educating, and Training Cadets and Junior Officers in the Modern World.* Elliot V. Converse, ed. Chicago: Imprint Publications, 1998. 383–388.
Buruma, Ian. *Inventing Japan: 1853–1964.* New York: Modern Library, 2003.
Campaign Planning Primer AY 06. U.S. Army War College. Accessed 19 June 2006. Available at http://www.carlisle.army.mil/usawc/dmspo/Publications/Publications1.htm.
Capstone Concept for Joint Operations. August 2005. Dept of Defense. Available at http://www.dtic.mil/futurejointwarfare/concepts/approved_ccjov2.pdf. Accessed 19 June 2006.
Carafano, James Jay and Alane Kochems. "Rethinking Professional Military Education." *The Heritage Foundation.* 28 July 2005. Available at www.heritage.org/research/nationalsecurity/em976.cfm.
Central Intelligence Agency. *The World Factbook.* Available at http://www.cia.gov/cia/publications/factbook/.
Chalfont, Alun. *Montgomery of Alamein.* New York: Atheneum, 1976.
Chamberlain, Peter, and Hilary Doyle. *Encyclopedia of German Tanks of World War Two.* London: Arms & Armour Press, 2001.
———, and Chris Ellis. *British & American Tanks of World War Two.* London: Arms & Armour Press, 2002.
Chuikov, Vasili I. *The Battle for Stalingrad.* Harold Silver, trans. New York: Holt, Rinehart & Winston, 1964.
Churchill, Winston S. *The Second World War.* 6 vols. Boston: Houghton Mifflin, 1983.
Clark, Alan. *Barbarossa: The Russian-German Conflict, 1941–1945.* New York: William Morrow & Co., 1965.
Clausewitz, Carl von. *On War.* Michael Howard & Peter Paret, eds. & trans. Princeton, NJ: Princeton Univ. P., 1976.
Cohen, Eliot. "Neither Fools Nor Cowards." *The Wall Street Journal.* 13 May 2005.
Colin, Jean. *The Transformation of War.* London: Hugh Rees, 1912.

Condon, Richard W. *The Winter War: Russia Against Finland.* New York: Ballantine, 1972.
Conquest, Robert. *The Great Terror: A Reassessment.* New York: Oxford University Press, 1990.
Cooper, James Fenimore. *The American Democrat.* Indianapolis: Liberty Fund, 1981.
Cooper, Matthew. *The German Army 1933–1945: Its Political & Military Failure.* New York: Stein & Day, 1978.
———. and James Lucas. *Panzer: The Armoured Force of the Third Reich.* New York: St. Martin's, 1978.
Craig, William. *Enemy at the Gates: The Battle for Stalingrad.* Old Saybrook, CT: Konecky & Konecky, 1973.
Crane, Conrad C., and W. Andrew Terrill. *Reconstructing Iraq: Challenges and Missions for Military Forces in a Post-Conflict Scenario.* Carlisle, PA: Strategic Studies Institute, 2003.
Creveld, Martin van. *The Training of Officers: From Military Professionalism to Irrelevance.* New York: Free Press, 1990.
Cru, Jean Norton. *War Books: A Study in Historical Criticism.* Stanley J. Pincetl and Ernst Marchand, eds. & trans. San Diego: San Diego State University Press, 1976.
De Gaulle, Charles. *The Army of the Future.* New York: J. B. Lippincott, 1941.
DeMontravel, Peter R. *A Hero to His Fighting Men: Nelson A. Miles, 1839–1925.* Kent, OH: Kent State University Press, 1998.
Derbyshire, John. "Apologizing for Iraq." 12 June 2006. *National Review Online.* Available at http://www.nationalreview.com.
———. "Chickenhawks (Series #947)." *National Review Online.* 29 March 2006. Available at http://www.nationalreview.com.
———. "Putting the World to Rights." *National Review Online.* 04 April 2006. Available at http://www.nationalreview.com.
———. "To Hell with the 'To Hell With Them Hawks' Hawks." *National Review Online.* 21 March 2006. Available at http://www.nationalreview.com.
———. "You Don't Know Jack." *National Review Online.* 09 March 2006. Available at http://www.nationalreview.com.
Dibold, Hans. *Doctor at Stalingrad.* Littleton: Aberdeen Bookstore, 2001.
Doyle, David. *The Standard Catalogue of German Military Vehicles.* Iola, WI: KP Books, 2005.
Draper, Theodore. *The Six Weeks' War.* New York: Viking, 1944.
Duffy, Christopher. *Red Storm Over the Reich.* Edison, NJ: Castle Books, 2002.
Dupuy, T. N. *A Genius For War.* Englewood Cliffs, NJ: Prentice-Hall, 1977.
Ellis, L. F. *The War in France & Flanders, 1939–1940.* London: Her Majesty's Stationery Office, 1953.
Elstob, Peter. *Condor Legion.* New York: Ballantine, 1973.
Erickson, John. *The Road to Stalingrad.* New York: Harper & Row, 1975.
Fallon, Daniel. *The German University.* Boulder: Colorado Associated University Press, 1980.

Farrar-Hockley, Anthony. *The War in the Desert.* London: Faber & Faber, 1969.
Foch, Ferdinand. *The Principles of War.* New York: H. K. Fly, 1918.
Ford, Brian. *German Secret Weapons: Blueprint for Mars.* New York: Ballantine, 1969.
Franks, Tommy. *American Soldier.* New York: ReganBooks, 2004.
Fraser, David. *Knight's Cross: A Life of Field Marshal Erwin Rommel.* New York: HarperCollins, 1993.
Fuller, J. F. C. *Decisive Battles of the Western World and Their Influence Upon History.* 3 vols. London: Cassell, 2001.
Gaddis, John Lewis. *Surprise, Security & the American Experience.* Cambridge, MA: Harvard University Press, 2004.
Gerlach, Heinrich. *The Forsaken Army.* London: Weidenfeld & Nicolson, 1958.
Gilbert, Martin. *Churchill: A Life.* New York: Henry Holt, 1991.
Giraudoux, Jean. *Tiger at the Gates.* Christopher Fry, trans. New York: Oxford University Press, 1955.
Glantz, David M. *Colossus Reborn: The Red Army at War, 1941–1943.* Lawrence, KS: University Press of Kansas, 2005.
Goerlitz, Walter. *History of the German General Staff, 1657–1945.* Boulder, CO: Praeger, 1985.
———. *Paulus & Stalingrad.* London: Methuen, 1963.
Gole, Henry G. *The Road to Rainbow: Army Planning for Global War, 1934–1940.* Annapolis: Naval Institute Press, 2003.
"Google to Censor Itself in China." *CNN.com.* 25 January 2006. Available at http://edition.cnn.com/2006/BUSINESS/01/25/google.china.
Gordon, Michael R., and Bernard E. Trainor. *Cobra II: The Inside Story of the Invasion and Occupation of Iraq.* New York: Pantheon, 2006.
Graves, Robert. *Goodbye to All That.* London: The Folio Society, 1981.
Gray, Colin. *Another Bloody Century: Future Warfare.* London: Weidenfeld & Nicolson, 2005.
Green, William. *Rocket Fighter.* New York: Ballantine, 1971.
Guderian, Heinz. *Achtung-Panzer!.* Christopher Duffy, trans. London: Arms & Armour Press, 1992.
Hart, Basil Liddell. *The History of the Second World War.* London: Cassell, 1970.
———. *The Remaking of Modern Armies.* Boston: Little, Brown & Co., 1928.
———, ed. *The Soviet Army.* London: Weidenfeld & Nicolson, 1956.
———. *Strategy: The Indirect Approach.* New York: Praeger, 1954.
———. *Why Don't We Learn From History?* New York: Hawthorn Books, 1971.
Hastings, Max. *Armageddon: The Battle for Germany 1944–1945.* New York: Knopf, 2004.
———. *Das Reich: Resistance & the March of the 2nd SS Panzer Division Through France, June 1944.* New York: Henry Holt, 1982.
Hayward, Joel. *Stopped at Stalingrad: The Luftwaffe & Hitler's Defeat in the East, 1942–1943.* Lawrence, KS: University Press of Kansas, 2004.
Hinz, Rolf. *To the Bitter End: The Final Battles of Army Groups North Ukraine, A, Centre, Eastern Front 1944–1945.* Frederick P. Steinhardt, trans. Solihull: Helion & Co., 2005.
Hittle, J. D. *The Military Staff: Its History & Development.* Harrisburg, PA: Military Service Publishing Co., 1949.

Hogg, Ian V. *Barrage: The Guns in Action.* New York: Ballantine, 1970.
———. *German Artillery of World War Two.* London: Greenhill Books, 1997.
———. *Grenades and Mortars.* New York: Ballantine, 1974.
———. *The Guns 1939–1945.* New York: Ballantine, 1970.
Holmes, Richard. *Bir Hakim: Desert Citadel.* New York: Ballantine, 1971.
Horne, Alistair. *The Price of Glory: Verdun 1916.* New York: St. Martin's, 1963.
———. *To Lose a Battle: France 1940.* Boston: Little, Brown & Co., 1969.
Horrocks, Brian. *A Full Life.* London: Collins, 1960.
Howard, Michael. *The Franco-Prussian War: The German Invasion of France, 1870–1871.* Norwalk, CT: Easton Press, 1994.
———. *The Invention of Peace.* New Haven, CT: Yale University Press, 2001.
———. *The Lessons of History.* New Haven: Yale University Press, 1991.
———, ed. *The Theory & Practice of War.* London: Cassell, 1965.
———. *War in European History.* New York: Oxford University Press, 1976.
Hughes, Daniel J., ed. *Moltke on the Art of War: Selected Writings.* Novato, CA: Presidio, 1993.
Humble, Richard. *Hitler's High Seas Fleet.* New York: Ballantine, 1971.
Huntington, Samuel P. *Who Are We? The Challenges to America's National Identity.* New York: Simon & Schuster, 2004.
Jessup, Philip C. *Elihu Root.* New York: Dodd, Mead & Co., 1938.
Johnson, David E. *Fast Tanks & Heavy Bombers: Innovation in the U.S. Army, 1917–1945.* Ithaca, NY: Cornell University Press, 1998.
Johnson, Paul. *A History of the American People.* New York: HarperCollins, 1997.
Joint Forces Staff College Pub 1: The Joint Staff Officer's Guide 2000.
Joint Publication 5-0: Doctrine for Planning Joint Operations. April 2006.
Jomini, Antoine-Henri. *The Art of War.* Charles Messenger, trans. London: Greenhill, 1996.
Jukes, Geoffrey. *Stalingrad.* New York: Ballantine, 1968.
Junger, Ernst. *The Storm of Steel.* New York: Howard Fertig, Inc., 1996.
Keegan, John. *Barbarossa: Invasion of Russia 1941.* New York: Ballantine, 1970.
———. *The Face of Battle.* New York: Viking, 1976.
———. *The First World War.* New York: Knopf, 1999.
———. *A History of Warfare.* New York: Knopf, 1993.
———. *Intelligence in War.* New York: Knopf, 2003.
———. *The Iraq War.* New York: Knopf, 2004.
———. *The Second World War.* New York: Penguin, 1990.
Kennedy, Paul M., ed. *The War Plans of the Great Powers, 1880–1914.* London: George Allen & Unwin, 1979.
Kershaw, Robert J. *War Without Garlands: Operation Barbarossa 1941/1942.* New York: Sarpedon, 2000.
Kier, Elizabeth. "Culture and Military Doctrine: France between the Wars." *International Security,* Spring 1995, 65–93.
Kipling, Rudyard. *Barrack Room Ballads.* New York: Alex Grosset & Co., 1899.
Kissinger, Henry. *Diplomacy.* New York: Simon & Schuster, 1994.
———. *A World Restored: Metternich, Castlereagh, and the Problems of Peace 1812–1822.* London: Weidenfeld & Nicolson, 1999.

Kurtz, Stanley. *Weekly Standard.* "Polygamy vs. Democracy: You Can't Have Both." *Weekly Standard,* 5 June 2006. 18–27.
Leach, Barry. *The German General Staff.* New York, Ballantine, 1973.
LeSage, Laurent. *Jean Giraudoux: His Life & Works.* State College, PA: Pennsylvania State University Press, 1959.
Lewis, Cecil. *Sagittarius Rising.* London: The Folio Society, 1998.
Lucas, James. *Storming Eagles: German Airborne Forces in World War Two.* London: Arms & Armour, 1988.
———. *War On the Eastern Front 1941–1945.* New York: Crown, 1979.
Lukacs, John. *Remembered Past: On History, Historians, and Historical Knowledge.* Wilmington, DE: ISI Books, 2005.
Luvaas, Jay, ed. & trans. *Frederick the Great on the Art of War.* New York: Da Capo Press, 1999.
MacGregor, Douglas A. *Transformation Under Fire: Revolutionizing How America Fights.* Westport, CT: Praeger, 2003.
Macksey, Kenneth. *Afrika Korps.* New York: Ballantine, 1968.
———. *Beda Fomm: The Classic Victory.* New York: Ballantine, 1971.
———. *From Triumph to Disaster: The Fatal Flaws of German Generalship from Moltke to Guderian.* Mechanicsburg, PA: Stackpole Books, 1996.
———. *Guderian: Panzer General.* London: Greenhill Books, 2003.
———. *Panzer Division: The Mailed Fist.* New York: Ballantine, 1968.
———. *The Shadow of Vimy Ridge.* Toronto: Ryerson, 1965.
———. *Tank Force: Allied Armor in World War II.* New York: Ballantine, 1970.
Manstein, Erich von. *Lost Victories.* Chicago: Regnery, 1958.
McCarthy, Andrew C. "The Way It Is." 22 March 2006. *National Review Online.* Available at http://www.nationalreview.com.
McCarthy, Peter, & Mike Syron. *Panzerkrieg: The Rise & Fall of Hitler's Tank Divisions.* New York: Carroll & Graf, 2002.
McClelland, Charles E. *State, Society, & University in Germany 1700–1914.* Cambridge: Cambridge University Press, 1980.
Milsom, John. *Russian Tanks 1900–1970: The Complete Illustrated History of Soviet Armoured Theory & Design.* Harrisburg, PA: Stackpole Books, 1971.
Mitcham, Samuel W. *Hitler's Legions: The German Army Order of Battle, World War II.* New York: Stein & Day, 1985.
———. *The Panzer Legions.* Westport, CT: Greenwood, 2001.
Motlke, Helmuth von. *The Franco-Prussian War of 1870–1871.* London: Greenhill Books, 1992.
Monroe, James. "The Monroe Doctrine." Rptd. in vol. 5 of *The Annals of America.* Chicago: Encyclopedia Britannica, 1976.
Morison, Samuel Eliot. *The Oxford History of the American People.* New York: Oxford University Press, 1965.
The National Defense Strategy of the United States of America. March 2005. Available at http://www.defenselink.mil/pubs/.
The National Military Strategy of the United States of America. 2004. Available at http://www.defenselink.mil/pubs/.
The National Security Strategy of the United States of America. March 2006. Available at http://www.whitehouse.gov/nsc/nss/2006/.

The National Strategy for Victory in Iraq. November 2005. Available at http://www.whitehouse.gov/infocus/iraq/iraq_national_strategy_20051130.pdf.

Neillands, Robin. *Eighth Army: From the Western Desert to the Alps, 1939–1945.* London: John Murray, 2004.

Netanyahu, Benjamin. *Fighting Terrorism.* New York: Farrar, 2001.

Office of Force Transformation. *Military Transformation: A Strategic Approach.* Available at http://www.oft.osd.mil/. Accessed 15 April 2006.

O'Hanlon, Michael E. "Iraq Without a Plan." *Policy Review.* Dec 04/Jan 05. Available at http://www.policyreview.org/dec04/ohanlon.html.

Orgill, Douglas. *German Armor.* New York: Ballantine, 1974.

———. *T-34: Russian Armor.* New York: Ballantine, 1971.

Ortega y Gasset, Jose. *The Revolt of the Masses.* New York: W. W. Norton, 1932.

Owens, Mackubin Thomas. "Rumsfeld & His Critics." *The Weekly Standard.* 03 May 2006. Available at http://www.weeklystandard.com.

Paret, Peter. *Understanding War.* Princeton: Princeton University Press, 1992.

Patton, George S. *War as I Knew It.* Boston: Houghton Mifflin, 1947.

Perret, Geoffrey. *Eisenhower.* New York: Random House, 1999.

Perrett, Bryan. *Knights of the Black Cross: Hitler's Panzerwaffe & Its Leaders.* New York: St. Martin's, 1986.

Plievier, Theodor. *Stalingrad.* Richard & Clara Winston, trans. New York: Appleton Century-Crofts, Inc., 1948.

Postgate, Raymond. *Story of a Year: 1848.* New York: Oxford University Press, 1956.

Price, Alfred. *Luftwaffe: Birth, Life & Death of an Air Force.* New York: Ballantine, 1969.

The Quadrennial Defense Review Report. 6 February 2006. Available at http://www.defenselink.mil/qdr/>.

Raus, Erhard. *Panzer Operations: The Eastern Front Memoir of General Raus, 1941–1945.* Steven H. Newton, comp. & trans. Cambridge, MA: Da Capo Press, 2003.

Reilly, John H. *Jean Giraudoux.* Boston: Twayne Publishers, 1978.

Remarque, Erich. *All Quiet on the Western Front.* E. W. Wheen, trans. New York: Ballantine, 1991.

Reynolds, Nicholas E. *Basrah, Baghdad, & Beyond: The U.S. Marine Corps in the Second Iraq War.* Annapolis, MD: Naval Institute Press, 2005.

Ricks, Tom. "Army Historian Cites Lack of Postwar Plan for Iraq." *Washington Post,* 25 December 2004, A01.

Rommel, Erwin. *Infantry Attacks!.* Vienna, VA: Athena, 1979.

———. *The Rommel Papers.* B. H. Liddell Hart, ed. New York: Harcourt, Brace, and Co., 1953.

Ross, Steven T. *American War Plans 1890–1939.* London: Frank Cass, 2002.

——— ed. *U.S. War Plans 1938–1945.* Boulder, CO: Lynne Rienner, 2001.

Rumsfeld, Donald. "Beyond Nation Building." 14 February 2003. United States Department of Defense. Accessed 03 April 2006. Available at http://www.defenselink.mil.

Safire, William, ed. *Lend Me Your Ears: Great Speeches in History.* New York: Norton, 2004.

Scales, Robert H. "Studying the Art of War." *Washington Times*, 17 February 2005.
Schellendorff, Bronsart von. *The Duties of the General Staff*. London: Harrison & Sons, 1905.
Schneider, Wolfgang. *Tigers in Combat I*. Mechanicsburg, PA: Stackpole Books, 2004.
Schroter, Heinz. *Stalingrad*. New York: Dutton, 1958.
Seaton, Albert. *The Battle for Moscow 1941–1942*. New York: Stein & Day, 1971.
———. *The German Army 193–45*. New York: St. Martin's, 1982.
———. *The Russo-German War 1941–1945*. London: Arthur Barker, Ltd., 1971.
——— and Joan Seaton. *The Soviet Army 1918 to the Present*. New York: New American Library, 1986.
———. *Stalin As Military Commander*. New York: Praeger, 1976.
Seeckt, Hans. *Thoughts of a Soldier*. Gilbert Waterhouse, trans. London: Benn, 1930.
Shanker, Tom, and Eric Schmitt. "The Struggle for Iraq: The Military; Young Officers Join Debate Over Rumsfeld." *The New York Times*. 25 April 2006. Accessed 5 May 2006. Available at http://nytimes.com.
Sharansky, Natan. *The Case For Democracy: The Power of Freedom to Overcome Tyranny & Terror*. New York: Public Affairs, 2004.
Shirer, William. *The Collapse of the Third Republic: An Inquiry into the Fall of France in 1940*. New York: Simon & Schuster, 1969.
———. *The Rise & Fall of the Third Reich*. New York: Simon & Schuster, 1960.
Smerconish, Michael. "Listen to Lehman." *National Review Online*. 15 April 2004. Available at http://www.nationalreview.com.
Stiehm, Judith. *The U.S. Army War College: Military Education in a Democracy*. Philadelphia: Temple University Press, 2002.
Stockdale, James B. *Courage Under Fire: Testing Epictetus's Doctrines in a Laboratory of Human Behavior*. Stanford: Hoover Institute, 1993.
Sweetman, John. *Ploesti: Oil Strike*. New York: Ballantine, 1974.
Swinson, Arthur. *The Raiders: Desert Strike Force*. New York: Ballantine, 1968.
Thorwald, Jurgen. *Flight in the Winter: Russia Conquers, January to May, 1945*. Fred Wieck, ed. & trans. New York: Pantheon, 1951.
Tieke, Wilhelm. *The Caucasus & the Oil: The German-Soviet War in the Caucasus 1942/1943*. Joseph G. Welsh, trans. Winnipeg: J. J. Fedorowicz Publishing, 1995.
Tocqueville, Alexis de. *Democracy in America*. Harvey C. Mansfield, trans. & ed. Chicago: University of Chicago Press, 2000.
Trevor-Roper, Hugh. *Blitzkrieg to Defeat: Hitler's War Directives*. New York: Holt, Rinehart & Winston, 1964.
Trotter, William R. *A Frozen Hell: The Russo-Finnish Winter War of 1939–1940*. Chapel Hill, NC: Algonquin Books, 1991.
Tuchman, Barbara. *The Guns of August*. London: Folio Society, 1995.
———. *The March of Folly*. New York: Knopf, 1984.
———. *The Proud Tower*. London: Folio Society, 1995.
———. *The Zimmermann Telegram*. London: Folio Society, 2004.

Webb, James. "Heading for Trouble: Do We Really Want to Occupy Iraq for the Next 30 Years?" *Jameswebb.com*. 4 September 2002. Accessed 20 May 2006. Available at http://www.jameswebb.com>.

Weber, Max. *On Universities: The Power of the State & the Dignity of the Academic Calling in Imperial Germany*. Edward Shils, trans. & ed. Chicago: University of Chicago Press, 1974.

Weeks, John. *Infantry Weapons*. New York: Ballantine, 1971.

———. *Men Against Tanks: A History of Antitank Warfare*. New York: Mason/Charter, 1975.

Wieder, Joachim, & Heinrich Graf von Einsiedel. *Stalingrad: Memories & Reassessments*. Helmut Bogler, trans. London: Arms & Armour Press, 1993.

Weigley, Russell F. *History of the United States Army*. New York: Macmillan, 1967.

Wilkinson, Spenser. *The Brain of an Army*. London: Constable & Co., 1913.

Will, George. "On Bilingual Ballots." 25 May 2006. Available at http://www.townhall.com.

Williams, John. *France: Summer 1940*. New York: Ballantine, 1970.

The World at War. Vols. 2 & 3. Prod., Jeremy Isaacs. Lawrence Olivier, narrator. 30th Anniversary Edition. A&E Home Video, 2004.

WWIIvehicles.com. http://www.wwiivehicles.com/index. Accessed 19 June 2006.

Zabecki, David T. "Germany's Forgotten Panzer Commander: Hermann Balck." Accessed 1 June 2006. Available at http://www.11thpanzer.com/dsp_balck.htm.

Zhukov, Georgi K. *Marshal Zhukov's Greatest Battles*. New York: Cooper Square Press, 2002.

Ziemke, Earl F. *The U.S. Army in the Occupation of Germany, 1944–1946*. Washington, D.C.: U.S. Army Center for Military History, 1990. Available at http://www.army.mil/cmhpg/books/wwii/Occ-GY/.

Zinni, Tony. *The Battle For Peace: A Frontline Vision of America's Power & Purpose*. New York: Palgrave Macmillan, 2006.

Index

Academic freedom, 98–99, 102–103
Achtung!-Panzer! (Guderian), 134
Adenauer, Konrad, 150
Admission requirements, academic, 85–86, 88–89, 103–104
Adolphus, Gustaphus, 4
Afghanistan, 23–24
Alexander, Harold, 175, 178
All Quiet on the Western Front (Remarque), 122
All the World's Airships, 112
All the World's Fighting Ships, 112
American Commonwealth, The (Bryce), 2, 41
American Democrat, The (Cooper), 41
American Soldier (Franks), 20–21
Army of the Future, The (de Gaulle), 127–130
Art of War, The (Jomini), 4
Auchinleck, Claude, 178

Bagehot, Walter, 40, 110
Balance-of-power approach, 34, 49–50
Balck, Hermann, 158, 159

Ballantine Illustrated History of World II, xiii
Ballantine monographs, 112
Barbusse, Henri, 121–122
Barnett, Correlli, 179
Battle for Peace, The (Zinni), 34–35
Beevor, Antony, 146, 149, 160
Bernhardi, Friedrich von, 50
Bismark, Otto von, 118, 179
Blue Force Tracker, 20
Blum, Léon, 61
Bradley, Omar N., 175
Brain of an Army, The (Wilkinson), 84–85
Brauchitsch, Walther von, 136, 140
Breaking the Phalanx (Macgregor), 64
British General Staff, 85
Brooke, Alan, 179
Bryce, James, 2, 41
Bureaucracy, military, 6, 31
Burke, Edmund, 65
Buruma, Ian, 28
Bush, George W., 20

Caesar, Julius, 4
Campaign Planning, 98

Capabilities-based planning, 56, 59–61, 64. *See also* Planning, military
Capstone Concept for Joint Operations (CCJO), 8–9
CENTCOM, 19
Central Intelligence Agency (CIA), 39
Chamberlain, Neville, 61–62, 130
Chaos as strategy tool, 54–55
Charles XII, 4
Chesterton, G.K., 65
Christie, Walter, 70, 73–74
Churchill, Winston, 176–177, 182
Citizenship, changing views of, 41
Civil War, U.S., 3
Clark, Mark W., 175
Clausewitz, Carl von, 82, 104, 112
  on cultural factors in warfare, 63–64
  influence on U.S. military, 4
  on psychological factors in warfare, 63–64
  on troop levels, 25
  on war policies, 20, 92, 127
Cobra II (Gordon & Trainor), 17
Cold War, start of, 143–44
Colin, Jean, 112, 119
Continental Congress, 2–3
Cooper, James Fenimore, 41
Craig, William, 146, 149, 150, 159
Crane, Conrad C., 21
Cromwell, Oliver, 3
Crüwell, Ludwig, 179
Cultural/social factors in warfare, 11–12, 22–23, 37–42, 47–48, 83, 183
Cunningham, Alan, 179
Curriculum in military education, 90–91, 95–97, 101–102

Daladier, Édouard, 130
Decline and Fall of the Roman Empire, The (Gibbons), 104
de Gaulle, Charles, 127–130, 170
democracies, 37–43
Democracy in America (Tocqueville), 2, 40, 41

Department of Defense Dictionary, 44–45
Department of Defense (DOD), xv, 20, 44–45
Department of Homeland Security, 31
Derbyshire, John, 105
Desert Generals, The (Barnett), 179
Diplomacy (Kissinger), 48–49
Douhet, Giulio, 112
draft, military, 42–43
Dreyfus Affair, 126

Education of military leaders
  academic freedom, 98–99, 102–103
  admission requirements, 85–86, 88–89, 103–104
  curriculum, 90–91, 95–97, 101–102
  faculty, 87–89, 94–95, 103
  humanities emphasis, 104–107
  Joint Advanced Warfighting School, xv, 95–100
  Joint & Combined Warfighting School, xv
  Joint Forces Staff College, xv, 93–100, 151
  recommendations, xv–xvi, 100–104, 186
  students, 88–89
  weaknesses, xv, 12, 87–91, 93–94
Eisenhower, Dwight D., 78–79, 99, 175
Enemy at the Gates (Craig), 146
Enemy motivations, 9–12, 56
Europe-U.S. relations, 47–49

Faculty in military education, 87–89, 94–95, 103
Federal Bureau of Investigation, 31
Field experience, value of, 114–115
Finland, invasion by Russia, 1939, 13–15
Foch, Ferdinand, 119, 127
France
  attitudes towards war, 130, 139–140

Index

diplomatic weaknesses, 123–125
and Germany, 115–18, 135–140
literature, 120–122
military forces, 125–127
World War I, 119–20
World War II, 80–82, 127–130
See also War plans; Weapons
Franks, Tommy, 19–21, 35
Frederick the Great, 115
Fuller, J.F.C, 65, 144

Gaddis, John Lewis, 34
Galland, Adolph, 135
Gamelin, Maurice, 81–82, 124
Garibaldi, Italo, 174
German General Staff, 84–85, 131, 138
German War Academy, 85, 87
Germany
  and France, 115–118, 135–140
  literature, 122–123
  military leaders, 132–135, 157–168
  motivations of leaders, 140–141
  in North Africa, 173–175, 177–178
  postwar, 26–30, 131–133
  strategy mistakes, 142–143
  and United Kingdom, 142–143
  See also War plans; Weapons
Germany and the Next War (Bernhardi), 50
Gibbons, Edward, 104
Giraudoux, Jean, 120–121
Global war on terror (GWOT), 28–29, 31
Goebbels, Josef, 122
Goerlitz, Walter, 164–165
Gordon, Michael R., 17
Göring, Hermann, 72
Gray, Colin, 92
Guderian, Heinz, 71–72, 74, 91, 116, 164–165
Guerre De Troie N'Aura Pas Lieu, La (Giraudoux), 120–121

Halder, Franz, 136, 140, 165
Hanson, Victor Davis, 27

Harris, Sam, 47
Hart, Basil Liddel, 80
  on innovation, 65
  on military leadership, 54, 88, 160
  on warfare, 111–112, 131
Hawkins, Steve, 19
Heim, Ferdinand, 164
History, study of
  decline in, 43, 80
  distortion of facts, 65–66
  impact on author, xiii–xiv
  military leaders and, xiv–xv, 83, 185–186
  value, 79–80, 111–115, 168, 185–186
Hitler, Adolf: rise to power, 61–62, 133–134
  and Russia, 13–15, 117–118
  and Stalingrad invasion, 146–147, 150, 157
  and United Kingdom, 116–117
Horne, Alistair, 80, 124
Horrocks, Brian, 175–176
Howard, Michael, 65
Hume, David, 2
Hussein, Saddam, 16–17, 26, 29–30

Immigration policy reform, 37, 41–42
Implementation of Network-Centric Operations (NCO), 60
Industrial College of the Armed Forces, 6
Infantry Attacks! (Rommel), 142, 178
Infantry Journal, 99
Inventing Japan, 1853-1964 (Buruma), 28
Iran, threat of, 26
Iraq
  alternatives, 26–27, 29–31
  comparisons to Stalingrad invasion, 145–146, 151
  cultural issues, 22–23
  planning failures, 24–32
  postwar issues, 17–19, 21–23, 27, 30, 46–47
  reasons for invasion, 31–32
  troop levels, 25–27

Iraq War, The (Keegan), 17
Iraq Without a Plan (O'Hanlon), 22

Japan
  military transformation, 58
  postwar occupations, 26–29
Japanese Self-Defense Force (JSDF), 58
Joint Advanced Warfighting School (JAWS), xv, 95–100. *See also* Education of military leaders
Joint Chiefs of Staff, 18, 21
Joint & Combined Warfighting School, xv. *See also* Education of military leaders
Joint Forces Staff College (JFSC), xv, 93–100, 151. *See also* Education of military leaders
Joint Publication 5-0:Joint Operation Planning (JP 5-0), 9–12
Jomini, Henri de, 4
Jünger, Ernst, 122–123

Keegan, John, 17, 80, 126
Kellogg-Briand Act (1928), 55
Kennedy, Anthony, 43–44
Kier, Elizabeth, 81–82
Kipling, Rudyard, 126
Kissinger, Henry, 48–49
Kleist, Ewald von, 157–58, 180
Kluge, Hans Günther von, 140
Krushchev, Nikita, 145
Kurtz, Stanley, 38

Le Feu (Barbusse), 121–122
Lend-Lease Act, 117
Lewis, Cecil, 123
Leyser, Ernst, 148
Livingstone, Ken, 47–48

MacArthur, Douglas, 28, 46
Macgregor, Douglas A., 64–68, 70–72, 78–82
Mackensen, August von, 132
Macksey, Kenneth, 134
Mahan, Alfred Thayer, 3–4, 91, 104
Manstein, Erich von, 135, 137–138, 154, 160

Mattis, James, 15
Maunoury, Michel-Joseph, 53–54
Maurin, Joseph-Léon-Marie, 130
McCarthy, Peter, 159
Mein Kampf (Hitler), 61, 140
Miles, Nelson A., 4
Military chain of command conflicts, 76–77
Military draft, 42–43
Military education. *See* Education of military leaders
Military field experience, value of, 114–115
Military journals, 51–53, 89–90, 98–99, 108
Military leaders
  commanders, 162, 166–167
  critical thinking skills, xiv–xv, 111, 113–114
  failure to challenge civilian leaders, 21
  joint staff assignments, 107–110
  officership, xv
  staff officers, 115, 162
  See also Education of military leaders; Germany; United Kingdom; *individual names*
Military strategy: recommendations, 46–47, 49–50
  written as marketing tool, 50–53. *See also* Planning, military
Military Transformation, 59–64
Military transformation: authentic, 58–59, 186
  Iraq war as model, 16–19
  Japan, 58
  risks in, 83
  role of technology, 62–63
  Special Forces role, 17–18
  suggested need for, 58–61
  technology and, 17–18, 62–64
  and troop levels, 25–26
  United Kingdom, 58
  United States, 58–61
Military weapons. *See* Weapons
Model, Walter, 135
Moltke, Helmuth J.L. von, 109, 137

Moltke, Helmuth K.R. von, 4, 15–16, 55
Monaghan, Tom, 42
Monroe Doctrine, 33–34
Montgomery, Bernard, 78–79, 170, 175, 179, 181–184
Mussolini, Benito, 172–173, 175

Napoleon, 4, 118
National Defense Strategy, 32
National Defense University, 98
National Military Strategy (NMS), 32–33, 44–46, 49–50
National security policy, 31–34
National Security Presidential Directive 24, 20
National Security Strategy (NSS), 32–37, 44–45, 57
National Strategy for Victory in Iraq, The, 29–30
Nation-states, value of, 37
Naval War College, 3–4, 91
Nehring, Walther K., 179
Network-centric operations (NCO), 60, 146
North Korea, threat of, 26

O'Connor, Richard, 173, 176–177, 178
Office of Force Transformation (OFT), 58
Officers. *See* military leaders
O'Hanlon, Michael, 21
On War (Clausewitz), 4, 63–64, 92, 97
Operational planning, 5–7
Operation Availability, 18
Operation Iraqi Freedom (OIF), 15–27. *See also* Iraq
Ortega y Gasset, José, 128
Owens, Mackubin Thomas, 91

Patton, George, 11, 175
Paulus, Friedrich von, 147, 152, 160–168
Peel, Robert, 110
Perito, Robert, 19
Personnel policies, military, 12, 108–110, 162–163, 186

Pétain, Philippe, 126–127
Phillips, Melanie, 48
Planning, military
 culture/social aspects, 83, 112–113
 explained, 1, 26–27, 32–36, 44–46
 failures, 16, 24–32
 in Iraq, 17–19, 24–32
 limitations of resources, 77–78
 merger of operational/strategic, 7–8, 11–12, 144
 and military transformation, 17–18
 national plan recommendations, 46, 49–50
 postwar, 17–19, 21–23, 26–30, 46–47, 131–133
 See also Military strategy; War plans
Powell, Colin, 18
Principles of War, The (Foch), 119
Prittwitz, Maximilian, 179
Proceedings (United States Naval Institute), 90, 98
Promotion decisions. *See* Personnel policies, Military
Public opinion/debate on war, 10–11

Quadrennial Defense Review (QDR), 51–53, 55–57

Raeder, Erich, 133
RAND Corporation, 19
Raus, Erhard, 135, 159
Reconstructing Iraq (Crane & Terrill), 22–23
Reichenau, Walther von, 165
Remaking of Modern Armies, The (Hart), 88
Remarque, Erich, 122
Reserved Office Training Corps (ROTC), 104, 115
Revolt of the Masses, The (Ortega y Gasset), 128
Reynaud, Paul, 130
Reynolds, Nicholas E., 31
Reynolds v. United States, 38
Ricks, Tom, 19
Riggs, John M., 90

Ritchie, Neil, 170, 179
Rokossovsky, Konstantin, 155–156
Rommel, Erwin, 7, 116, 164
  characteristics of, 54–55, 170, 183–184
  memoir of, 142, 178
  in North Africa, 174, 177–181
Roosevelt, Franklin D., 41, 116–117
Root, Elihu, 4–5, 78, 85
Roper v. Simmons, 43–44
ROTC, 104, 115
Rousseau, Jean-Jacques, 2
Rumsfeld, Donald, 17–18, 19, 20, 91
Rundstedt, Karl von, 138
Russia
  in Eastern Europe, 143–144
  and Finland, 13–15
  leadership strengths, 155–156
  See also Weapons

Sagittarius Rising (Lewis), 123
Sartre, Jean-Paul, 120
Scharnhorst, Gerhard J., 85
Schlieffen, Alfred von, 27, 166
Schlieffen Plan, 1, 55, 135–136
Schmitt, Eric, 19
Scott, George, 54
Seeckt, Hans von, 132–133
Shanker, Tom, 19
Sherman tanks, 67–68, 175–176. *See also* Weapons
Shinseki, Eric, 19, 25, 27, 30
Smith, Adam, 2
Social/cultural factors in warfare, 11–12, 22–23, 37–42, 47–48, 83, 183
Speer, Albert, 152
Stalingrad (Beevor), 146
Stalingrad (film), 146
Stalingrad invasion. *See* World War II
Stalingrad (Wieder & Einsiedel), 160–161
Stalin, Joseph, 13–15, 143–144
Storm of Steel (Jünger), 122–123
Strategic planning. *See* Planning, military
Strategic Studies Institute, 21
Student, Kurt, 135

Stumme, Georg, 179
Systems approach to warfare, 8–10, 142

Tanks
  German, 66–67, 71–75, 171
  Russian, 66–71
  United States, 66–68, 171, 175–176
Technology
  emphasis in war plans, 8–10, 56, 111
  impact on warfare, 63–64
  role in military transformation, 62–64
Tedder, Arthur W., 175
Terrill, W. Andrew, 21
Thoma, Wilhelm von, 170, 179
threat-based planning, 56
Thucydides, 55
Time to Emigrate? (Walden), 48
Tocqueville, Alexis de, 2, 40, 41, 117
Tommy (Kipling), 126
Trainor, Bernard E., 17
Transformational diplomacy, 36
Transformation, military. *See* Military transformation
Transformation of War, The (Colin), 119
Transformation under Fire (Macgregor), 64–66, 78, 80–82
Treaty of Paris, 118
Trojan War Will Not Take Place, The (Giraudoux), 120–121
Troop levels, 25–27
Tuchman, Barbara, 24, 28–29, 32, 39
Turner, Stansfield, 90

Under Fire (Barbusse), 121–122
United Kingdom
  early military reforms, 84
  and Germany, 141–142
  and Iraq, 47
  military leaders, 178–179
  military transformation, 58
  in North Africa, 173, 175, 176–179
  and United States, 47–48
  See also War plans

Index

United States
  early military strategies, 3–4
  European nations relations,
    48–49
  post-WWII commitments, 27–29
  role changes since WWII, 34
  and United Kingdom, 47–48
  See also Iraq; Military
    transformation; War plans;
    Weapons
U.S. Air Force, xiii, 77
U.S. Armed Forces, 58
U.S. Army, 3, 5–6
U.S. Army War College, 6–7, 21
U.S. Central Command
  (CENTCOM), 19
U.S. Marines, 45
U.S. Naval Institute, 85
U.S. Navy, 3

Verecker, John S., 141–142
Versailles Treaty, 123–124
Vers l' Armée le Metier (de Gaulle),
  127–130
Volsky, Viktor, 148
Voltaire, 2

Walden, George, 48
War colleges, U.S. *See* Education of
  military leaders
Warfare
  American views of, 17–18
  culture/social aspects, 83,
    112–113
  defined, 45
  efficiency in, 17–18
  marketing of, 45–46
  as social intercourse, 112–113
  and technology, 17–18, 20, 63
  traditional practices, 18
  See also Planning, military;
    Weapons
Warfighting, 45
Warlimont, Walter, 180
War planning. *See* Planning, military
War plans
  France, 138–139

  German, 131–133, 135–138,
    151–152, 154–155
  United Kingdom, 173, 175–177
  United States, 6–7, 15–28, 76–78
  See also Planning, military;
    Schlieffen Plan
War Plans Division, 6–7
Washington, George, 2–3
Wavell, Archibald, 176, 177
Weapons
  France, 131
  Germany, 63, 71–76
  publications about, xiii, 112
  Russia, 68–70
  United States, 67–69, 77–78
  See also Tanks
Weapons of mass destruction
  (WMD), 26
Webb, James, 19, 30
Wegener, Wolfgang, 133
Weichs, Maximilian von, 166
Wellesley, Arthur, 4
Wieder, Joachim, 160–161
Wilkinson, Spenser, 84–85
Will, George, 41
William I, 118
Wilson, Isaiah, III, 19
Wilson, Woodrow, 6, 28, 33
World Factbook, The, 39
World War I, 28–29, 33, 53–54, 118
World War II
  Allies efforts, 78–79
  mistakes made, 80–82
  North Africa, 169–184
  postwar occupations, 26–30
  Scheldt Estuary, 78–79
  Stalingrad invasion, 145–168
  victory by attrition, 7
  See also War plans; Weapons
Worthington, George R., 90
Wykes, Alan, xiii

Zenker, Hans, 133
Zhukov, Georgy, 145
Ziemke, Earl F., 23, 29
Zimmermann, Arthur, 28–29
Zinni, Tony, 18–20, 25, 34–35

About the Author

BRIAN HANLEY serves as a member of the United States Naval Institute Editorial Board of Directors. His work has appeared in *Proceedings, Journal of Military History, Joint Forces Quarterly,* and elsewhere. He is the author of *Samuel Johnson as Book Reviewer,* which was published by the University of Delaware Press in 2001. A twenty-year veteran of the U.S. Air Force, Hanley holds advanced degrees in English from the University of Oxford and the University of North Carolina at Chapel Hill.